CAMBRIDGE LIBRARY COLLECTION

Books of enduring scholarly value

Physical Sciences

From ancient times, humans have tried to understand the workings of the world around them. The roots of modern physical science go back to the very earliest mechanical devices such as levers and rollers, the mixing of paints and dyes, and the importance of the heavenly bodies in early religious observance and navigation. The physical sciences as we know them today began to emerge as independent academic subjects during the early modern period, in the work of Newton and other 'natural philosophers', and numerous sub-disciplines developed during the centuries that followed. This part of the Cambridge Library Collection is devoted to landmark publications in this area which will be of interest to historians of science concerned with individual scientists, particular discoveries, and advances in scientific method, or with the establishment and development of scientific institutions around the world.

A New System of Chemical Philosophy

The chemist and meteorologist John Dalton (1766–1844) published *A New System of Chemical Philosophy* in two volumes, between 1808 and 1827. Dalton's discovery of the importance of the relative weight and structure of particles of a compound for explaining chemical reactions transformed atomic theory and laid the basis for much of what is modern chemistry. Volume 1 was published in two parts, in 1808 and 1810. Part 1 offers an account of Dalton's atomic theory. It contains chapters on temperature, the constitution of bodies, chemical synthesis and a number of plates including his famous table of symbols for the atoms of various elements. Part 2 contains a chapter on elementary principles and twelve sections on different groups of two-element compounds. Dalton's work is a monument of nineteenth-century chemistry. It will continue to be read and enjoyed by anybody interested in the history and development of science.

Cambridge University Press has long been a pioneer in the reissuing of out-of-print titles from its own backlist, producing digital reprints of books that are still sought after by scholars and students but could not be reprinted economically using traditional technology. The Cambridge Library Collection extends this activity to a wider range of books which are still of importance to researchers and professionals, either for the source material they contain, or as landmarks in the history of their academic discipline.

Drawing from the world-renowned collections in the Cambridge University Library, and guided by the advice of experts in each subject area, Cambridge University Press is using state-of-the-art scanning machines in its own Printing House to capture the content of each book selected for inclusion. The files are processed to give a consistently clear, crisp image, and the books finished to the high quality standard for which the Press is recognised around the world. The latest print-on-demand technology ensures that the books will remain available indefinitely, and that orders for single or multiple copies can quickly be supplied.

The Cambridge Library Collection will bring back to life books of enduring scholarly value (including out-of-copyright works originally issued by other publishers) across a wide range of disciplines in the humanities and social sciences and in science and technology.

A New System of Chemical Philosophy

VOLUME 1

JOHN DALTON

CAMBRIDGE UNIVERSITY PRESS

Cambridge, New York, Melbourne, Madrid, Cape Town, Singapore,
São Paolo, Delhi, Dubai, Tokyo

Published in the United States of America by Cambridge University Press, New York

www.cambridge.org
Information on this title: www.cambridge.org/9781108019675

This edition first published 1808-10
This digitally printed version 2010

ISBN 978-1-108-01967-5 Paperback

A NEW SYSTEM OF CHEMICAL PHILOSOPHY

PART I.

BY

JOHN DALTON.

Manchester:

Printed by S. Russell, 125, Deansgate,

FOR

R. BICKERSTAFF, STRAND, LONDON.

1808.

TO THE

PROFESSORS OF THE UNIVERSITIES,

AND OTHER RESIDENTS,

OF

EDINBURGH AND GLASGOW,

who gave

THEIR ATTENTION AND ENCOURAGEMENT

TO THE

Lectures on Heat and Chemical Elements,

Delivered in those Cities in 1807;

AND

TO THE MEMBERS OF THE

LITERARY AND PHILOSOPHICAL SOCIETY

OF MANCHESTER,

who have

UNIFORMLY PROMOTED HIS RESEARCHES;

THIS WORK IS RESPECTFULLY INSCRIBED,

BY THE

AUTHOR.

PREFACE.

IT was the author's intention when this work was put to press to publish it entire in one volume; but he is now induced to publish it in two parts, for reasons which it may be proper to announce.

Various essays of his were read before the Literary and Philosophical Society of Manchester, chiefly on heat and elastic fluids, and were published in the 5th Volume of their Memoirs, in 1802. The new views which these essays developed, were considered both curious and important. The essays were republished in several Philosophical Journals, and soon after translated into French and German, and circulated abroad through the medium of the foreign Journals. The author was not remiss in prosecuting his researches, in which he was considerably assisted by the application of principles derived from the above essays. In 1803, he was gradually led to those primary Laws, which seem to obtain in regard to heat, and to chemical combinations, and which it is the object of the present work to exhibit and elucidate. A brief outline of them was first publicly given the ensuing winter in a course of Lectures on Natural Philosophy, at the Royal Institution in London, and was left for publication in the Journals of the Institution; but he is not informed whether that was done. The author has ever since been occasionally urged by several of his philosophical friends to lose no time in communicating the results of his enquiries to the public, alledging, that the interests of science, and his own reputation, might suffer by delay. In the spring of 1807, he was induced to offer the exposition of the principles herein contained in a course of Lectures, which were twice read in Edinburgh, and once in Glasgow. On these occasions he was honoured with the attention of gentlemen, universally acknowledged to be of the first respectability for their scientific attainments: most of whom were pleased to express their desire to see the publication of the doctrine in the present form, as soon as convenient. Upon the author's return to Manchester he began to prepare for the press. Several experiments required to be repeated; other new ones were to be made; almost the whole system both in matter and manner was to be new, and consequently required more time for the composition and arrangement. These considerations, together with the daily avocations of profession, have delayed the work nearly a year; and, judging from the past, it may require another year before it can be completed. In the mean time, as the doctrine of heat, and the general principles of Chemical Synthesis, are in a good degree independent of the future details, there can no great detriment arise to the author, or inconvenience to his readers, in submitting what is already prepared, to the inspection of the public.

MAY, 1808.

CONTENTS OF PART FIRST.

NEW SYSTEM

OF

CHEMICAL PHILOSOPHY.

CHAP. I.

ON HEAT OR CALORIC.

THE most probable opinion concerning the nature of caloric, is, that of its being an elastic fluid of great subtilty, the particles of which repel one another, but are attracted by all other bodies.

When all surrounding bodies are of one temperature, then the heat attached to them is in a quiescent state; the absolute quantities of heat in any two bodies in this case are not equal, whether we take the bodies of equal weights or of equal bulks. Each kind of matter has its peculiar affinity for heat, by which it requires a certain portion of the fluid, in order to be in equilibrium with other bodies at a certain temperature. Were the *whole quantities* of heat in bodies of equal weight or bulk, or even the *relative quantities*, accurately ascertained, for any temperature, the numbers expressing those quantities would constitute a table of *specific heats*, analogous to a table of *specific gravities*, and would be an important acquisition to science. Attempts of this kind have been made with very considerable success.

Whether the specific heats, could they be thus obtained for one temperature, would express the relation at every other temperature, whilst the bodies retained their form, is an enquiry of some moment. From the experiments hitherto made there seems little doubt of its being nearly so; but it is perhaps more correct to deduce the specific heat of bodies from equal *bulks* than from equal *weights*. It is very certain that the two methods will not give precisely the same results, because the expansions of different bodies by equal increments of temperature are not the same. But before this subject can well be considered, we should first settle what is intended to be meant by the word temperature.

SECTION 1.

ON TEMPERATURE,

And the Instruments for measuring it.

The notion of the specific heat of bodies and of temperature, may be well conceived from a system of cylindrical vessels of different diameters connected with each other by pipes at the bottom, and a small cylindrical tube attached to the system, all capable of holding water or any other liquid, and placed perpendicular to the horizon. (See Plate 1. Fig. 1.) The cylinders are to represent the different specific heats of bodies; and the small tube, being divided into equal parts, is to represent the thermometer or measure of temperature. If water be poured into one vessel it rises to the same level in them all, and in the thermometer; if equal portions be successively poured in, there will be equal rises in the vessels and in the tube; the water is obviously intended to represent heat or caloric. According to this notion, then, it is evident that equal increments of heat in any body correspond to equal increments of temperature.

This view of the subject necessarily requires, that if two bodies be taken of any one temperature, and then be raised to any other temperature, the additional quantities of heat received by each will be exactly proportioned to the whole quantities of that fluid previously contained in them. This conclusion, though it may be nearly consistent with facts in general, is certainly not strictly true. For, in elastic fluids, it is well known, an increase of *bulk* occasions an increase of specific heat, though the weight and temperature continue the same. It is probable then that solids and liquids too, as they increase in bulk by heat, increase in their capacity or capability of receiving more. This circumstance, however, might not affect the conclusion above, provided all bodies increased in one and the same proportion by heat; but as this is not the case, the objection to the conclusion appears of validity. Suppose it were allowed that a thermometer ought to indicate the accession of *equal* increments of the fluid denominated caloric, to the body of which it was to shew the temperature;—suppose too that a measure of air or elastic fluid was to be the body; query, whether ought the air to be suffered to expand by the temperature, or to be confined to the same space of one measure? It appears to me the most likely in theory to procure a standard capacity for heat by subjecting a body to heat, *whilst its bulk is kept constantly the same.* Let m = the quantity of heat necessary to raise the elastic fluid 10° in temperature in this case; then $m + d$ = the quantity necessary to raise the same 10°, when suffered to expand, d being the difference of the absolute quantities of heat contained by the body in the two cases. Now, $\frac{1}{10} m$ = the quantity of heat necessary to raise the temperature 1° in the first case; but $\frac{1}{10}(m+d)$ can not be the quantity necessary in the second case; it will be a less quantity in the lower degrees, and a greater in the higher. If these principles be admitted, they may be applied to liquids and solids; a liquid, as water, cannot be raised in temperature equally by equal increments of heat, unless it is confined within the same space by an extraordinary and perhaps incalculable force; if we suffer it to take its ordinary course of expansion, then, not equal, but increasing increments of heat will raise its temperature uniformly. If sufficient force were applied to condense a liquid or solid, there can be no doubt but heat would be given out, as with elastic fluids.

It may perhaps be urged by some that the difference of heat in condensed and rarefied air, and by analogy probably in the supposed cases of liquids and solids, is too small to have sensible influence on the capacities or affinities of bodies for heat; that the effects are such, as only to raise or depress the temperature a few degrees; when perhaps the whole mass of heat is equivalent to two or three thousand such degrees; and that a volume of air supposed to contain 2005° of temperature being rarefied till it become 2000°, or lost 5° of temperature, may still be considered as having its capacity invariable. This may be granted if the data are admissible; but the true changes of temperature consequent to the condensation and rarefaction of air have never been determined. I have shewn, (Manchester Mem. Vol. 5, Pt. 2.) that in the process of admitting air into a vacuum, and of liberating condensed air, the inclosed thermometer is affected as if in a medium of 50° higher or lower temperature; but the effects of instantaneously doubling the density of air, or replenishing a vacuum, cannot easily be derived from those or any other facts I am acquainted with; they may perhaps raise the temperature one hundred degrees or more. The great heat produced in charging an air-gun is a proof of a great change of capacity in the inclosed air.— Upon the whole then it may be concluded, that the change of bulk in the same body by change of temperature, is productive of considerable effect on its capacity for heat, but that we are not yet in possession of data to determine its effect on elastic fluids, and still less on liquids and solids. M. De Luc found, that in mixing *equal weights* of water at the freezing and boiling temperatures, 32° and 212°, the mixture indicated nearly 119° of Fahrenheit's mercurial thermometer; but the numerical mean is 122°; if he had mixed *equal bulks* of water at 32° and 212°, he would have found a mean of 115° Now the means determined by experiment in both these ways are probably too high; for, water of these two temperatures being mixed, loses about 1-90th of its bulk; this condensation of volume (whether arising from an increased affinity of aggregation, or the effect of external mechanical compression, is all one) must expel a quantity of heat, and raise the temperature above the true mean. It is not improbable that the true mean temperature between 32° and 212° may be as low as 110° of Fahrenheit.

It has been generally admitted that if two portions of any liquid, of equal weight but of different temperatures, be mixed together, the mixture must indicate the true mean temperature; and that instrument which corresponds with it is an accurate measure of temperature. But if the preceding observations be correct, it may be questioned whether any two liquids will agree in giving the same mean temperature upon being mixed as above.

In the present imperfect mode of estimating temperature, the equable expansion of mercury is adopted as a scale for its measure. This cannot be correct for two reasons; 1st. the mixture of water of different temperatures is always *below* the mean by the mercurial thermometer; for instance, water of 32° and 212° being mixed, gives 119° by the thermometer; whereas it appears from the preceding remarks, that the temperature of such mixture ought to be found above the mean 122°; 2d. mercury appears by the most recent experiments to expand by the same law as water; namely, as the square of the temperature from the point of greatest density.—The apparently equal expansion of mercury arises from our taking a small portion of the scale of expansion, and that at some distance from the freezing point of the liquid.

From what has been remarked it appears that we have not yet any mode easily practicable for ascertaining what is the true mean between any two temperatures, as those of freezing and boiling water; nor any thermometer which can be considered as approximating nearly to accuracy.

Heat is a very important agent in nature; it

cannot be doubted that so active a principle must be subject to general laws. If the phenomena indicate otherwise, it is because we do not take a sufficiently comprehensive view of them. Philosophers have sought, but in vain, for a body that should expand uniformly, or in arithmetical progression, by equal increments of heat; liquids have been tried, and found to expand unequally, all of them expanding more in the higher temperatures than in the lower, but no two exactly alike. Mercury has appeared to have the least variation, or approach nearest to uniform expansion, and on that and other accounts has been generally preferred in the construction of thermometers. Water has been rejected, as the most unequally expanding liquid yet known. Since the publication of my experiments on the expansion of elastic fluids by heat, and those of Gay Lussac, immediately succeeding them, both demonstrating the perfect sameness in all permanently elastic fluids in this respect; it has been imagined by some that gases expand equally; but this is not corroborated by experience from other sources.

Some time ago it occurred to me as probable, that water and mercury, notwithstanding their apparent diversity, actually expand by the same law, and that the quantity of expansion is as the square of the temperature from their respective freezing points. Water very nearly accords with this law according to the present scale of temperature, and the little deviation observable is exactly of the sort that ought to exist, from the known error of the equal division of the mercurial scale. By prosecuting this enquiry I found that the mercurial and water scales divided according to the principle just mentioned, would perfectly accord, as far as they were comparable; and that the law will probably extend to all other pure liquids; but not to heterogeneous compounds, as liquid solutions of salts.

If the law of the expansion of liquids be such as just mentioned, it is natural to expect that other phenomena of heat will be characteristic of the same law. It may be seen in my Essay on the Force of Steam (Man. Mem. Vol. 5, Part 2.) that the elastic force or tension of steam in contact with water, increases *nearly* in a geometrical progression to equal increments of temperature, *as measured by the common mercurial scale*; it was not a little surprising to me at the time to find such an approach to a regular progression, and I was then inclined to think, that the want of perfect coincidence was owing to inaccuracy in the division of the received thermometer; but overawed by the authority of Crawford, who seemed to have proved past doubt that the error of the thermometer no where amounted to more than one or two degrees, I durst not venture to throw out more than a suspicion at the conclusion of the essay, on the expansion of elastic fluids by heat, that the error was probably 3 or 4°, as De Luc had determined; to admit of an error in the supposed mean, amounting to 12°, seemed unwarrantable. However it now appears that the force of steam in contact with water, increases *accurately* in geometrical progression to equal increments of temperature, provided those increments are measured by a thermometer of water or mercury, the scales of which are divided according to the above-mentioned law.

The Force of Steam having been found to vary by the above law, it was natural to expect that of air to do the same; for, air (meaning any permanently elastic fluid) and steam are essentially the same, differing only in certain modifications. Accordingly it was found upon trial that air expands in geometrical progression to equal increments of temperature, measured as above. Steam detached from water, by which it is rendered incapable of increase or diminution in quantity, was found by Gay Lussac, to have the same quantity of expansion as the permanently elastic fluids. I had formerly conjectured that air expands as the *cube* of the temperature from absolute privation, as hinted in the essay above-mentioned; but I am now obliged to abandon that conjecture.

The union of so many analogies in favour the preceding hypothesis of temperature is almost sufficient to establish it; but one remarkable trait of temperature derived from experiments on the heating and cooling of bodies, which does not accord with the received scale, and which, nevertheless, claims special consideration, is, that *a body in cooling loses heat in proportion to its excess of temperature above that of the cooling medium*; or that the temperature descends in geometrical progression in equal moments of time. Thus if a body were 1000° above the medium; the times in cooling from 1000° to 100, from 100 to 10, and from 10 to 1°, ought all to be the same. This, though nearly, is not accurately true, if we adopt the common scale, as is well known; the times in the lower intervals of temperature are found longer than in the upper; but the new scale proposed, by shortening the lower degrees, and lengthening the higher, is found perfectly according to this remarkable law of heat.

Temperature then will be found to have four most remarkable analogies to support it.

1st. All pure homogenous liquids, as water and mercury, expand from the point of their congelation, or greatest density, a quantity always as the square of the temperature from that point.

2. The force of steam from pure liquids, as water, ether, &c. constitutes a geometrical progression to increments of temperature in arithmetical progression.

3. The expansion of permanent elastic fluids is in geometrical progression to equal increments of temperature.

4. The refrigeration of bodies is in geometrical progression in equal increments of time.

A mercurial thermometer graduated according to this principle will differ from the ordinary one with equidifferential scale, by having its lower degrees smaller and the upper ones larger; the mean between freezing and boiling water, or 122° on the new scale, will be found about 110° on the old one.—The following Table exhibits the numerical calculations illustrative of the principles inculcated above.

NEW TABLE OF TEMPERATURE.

	Mercury.				*Water*	*Air.*	*Vapour.*		
True equal intervals of temperature.	Roots, or intervals of temperature, com. dif =.4105	Squares, or measures of temp. on merc. scale frm freezing merc.	Same as preceding column, —40° or Farenheit's scale.	Common Farenheit's scale; or preceding colum corrected for expansion of glass.	Expansion of water as square of temp	Expansion of air in geomet. progress Ratio 1,0179	Force of vapour of water geom. prog ratio 1,321. In. M	Force of vapour of ether, geom prog. ratio 1.2278 Inch M.	Force of vapour from alcohol irregular Sp. Gr. ,87 In. M.
—175°	0 —	0—	—40°—	——	——	692 —			
— 68°	4.3803	18.88	—21.12			837.6	.012	.78	
— 58	4 7908	22.94	—17.06			852.5	.016	.96	
— 48	5.2013	27 04	—12 96			867.7	.022	1.18	
— 38	5.6118	31 58	— 8.52			883 3	.028	1.45	
— 28	6.0223	36 24	— 3.76			899.—	.088	1 37	
— 18	6.4328	41.34	1 34			915.2	.050	2.17	.47
— 8	6 8433	46.78	6.78			931.5	.066	2.68	.52
2	7.2538	52.63	12 63		16	948.2	.087	3 30	.58
12	7 6643	58 74	18.74		9	965.2	.115	4,05	.64
22	8.0748	65 21	25 21		4	982.4	.151	4.07	.72
32	8 4853	72.—	32.—	32°	1	1000	.200	6.1	.80
42	8.8958	79.1	39.1	39.3	0	1017.9	.264	7.57	.93
52	9 3063	86.6	46.6	47.—	1	1036.1	.348	9.16	1.08
62	9.7108	94.44	54 44	55 —	4	1054 7	.461	11.22	1.3
72	10.1273	102.55	62 55	63.3	9	1073.5	.609	13.77	1.6
82	10.5378	111.04	71.04	72.—	16	1092.7	,804	16.85	2.1
92	10.9483	119.84	79 84	81	25	1112.3	1.062	20,65	2 8
102	11.3588	129 02	89 02	90.4	36	1132 2	1.40	25.30	3 6
112	11.7693	138.49	98 49	100.1	49	1152 4	1,85	31.—	4.7
122	12.1798	148 3	108.3	110.—	64	1173.1	2 46	37 98	6.3
132	12.5903	158 5	118.5	120.1	81	1194 —	3.24	46 54	8.2
142	13 0008	169 —	129.—	130 4	100	1215.4	4.27	57.03	10.2
152	13 4113	179 9	139.9	141.1	121	1237.1	5.65	69 88	13.9
162	13.8218	191.—	151 —	152 —	144	1259.2	7.47	85.62	17.9
172	14.2323	202.4	162.4	163 2	169	1281 8	9 87	104.91	22 4
182	14.6428	214.4	174 4	175 —	196	1304 7	13.02	128.5	29.3
192	15.0533	226.5	186.5	186 9	225	1328.	17.19	157 5	
202	15.4638	239	199.—	199 2	256	1351 8	22.70	193.—	
212	15.8743	252	212.	212 —	289	1376.—	30 00	236.5	
312	19.9793	399.1	359 1			1643	485.—		
412	24 0843	579 8	539 8			1962			
512	28.1893	794.7	754 7			2342			
612	32.2943	1043.—	1000.			2797			
712	336.399	1325.—	1285.			3339			

Explanation of the Table.

The first column contains the degrees of temperature, of which there are supposed to be 180 between freezing and boiling water, according to Fahrenheit. The concurrence of so many analogies as have been mentioned, as well as experience, indicate that those degrees are produced by equal increments of the matter of heat, or caloric; but then it should be understood they are to be applied to a body of uniform bulk and capacity, such as air confined within a given space. If water, for instance, in its ordinary state, is to be raised successively through equal intervals of temperature, as measured by this scale, then unequal increments of heat will be requisite, by reason of its increased capacity. The first number in the column, —175°, denotes the point at which mercury freezes, hitherto marked —40°. The calculations are made for every 10° from —68° to 212°; above the last number, for every 100°. By comparing this column with the 5th, the correspondences of the new scale and the common one are perceived: the greatest difference between 32° and 212° is observable at 122° of the new scale, which agrees with 110° of the old, the difference being 12°; but below 32° and above 212°, the differences become more remarkable.

The 2d and 3d columns are two series, the one of roots, and the other of their squares. They are obtained thus; opposite 32°, in the first column, is placed in the 3d, 72°, being the number of degrees or equal parts in Fahrenheit's scale from freezing mercury to freezing water; and opposite 212° in the first is placed 252° in the 3d, being 212 + 40°, the number of degrees (or rather equal parts) between freezing mercury and boiling water. The square roots of these two numbers, 72° and 252°, are found and placed opposite to them in the second column. The number 8.4853 represents the relative quantity of real temperature between freezing mercury and freezing water; and the number 15.8743 represents the like between freezing mercury and boiling water; consequently the difference 7.3890 represents the relative quantity between freezing water and boiling water, and 7.3890 ÷ 18 = .4105 represents the quantity corresponding to each interval of 10°. By adding .4105 successively to 8.4853, or subtracting it from it, the rest of the numbers in the column are obtained, which are of course in arithmetical progression. The numbers in the 3d column are all obtained by squaring those of the 2d opposite to them. The unequal differences in the 3d column mark the expansions of mercury due to equal increments of temperature, by the theory. The inconvenient length of the table prevents its being carried down by intervals of 10° to the point of freezing mercury, which however is found to be at —175°.

The 4th column is the same as the 3d, with the difference of 40°, to make it conform to the common method of numbering on Fahrenheit's scale.

The 5th column is the 4th corrected, on account of the unequal expansion of Glass:— The *apparent* expansion of mercury in glass is less than the *real*, by the expansion of the glass itself; this, however, would not disturb the law of expansion of the liquid, both apparent and real being subject to the same, *provided the glass expands equally*; this will be shewn hereafter. But it has been shewn by De Luc, that glass expands less in the lower half of the scale than the higher; this must occasion the mercury apparently to expand more in the lower half than what is dictated by the law of expansion. By calculating from De Luc's data, I find, that the mercury in the middle of the scale, or 122°, ought to be found nearly 3° higher than would be, were it not for this increase. Not however to over-rate the effect, I have taken it only at 1°.7, making the number 108°, 3 in the 4th column, 110° in the 5th, and the rest of the column is corrected accordingly. The numbers in this column cannot well be extended much beyond the interval from freezing to boiling water, for want of experiments on the expansion of glass. By viewing this column along with the 1st, the quantity of the supposed error in the common scale may be perceived; and any observations on the old thermometer may be reduced to the new.

The 6th column contains the squares of the natural series 1, 2, 3, &c. representing the expansion of water by equal intervals of temperature. Thus, if a portion of water at 42° expands a quantity represented by 289, at the boiling temperature, then at 52° it will be found to have expanded 1, at 62°, 4 parts, &c. &c. Water expands by cold or the abstraction of heat in the same way below the point of greatest density, as will be illustrated when we come to consider the absolute expansion of bodies. The apparent greatest density too does not happen at 39°,3 old scale; but about 42°; and the greatest real density is at or near 36° of the same.

The 7th column contains a series of numbers in Geometrical Progression, denoting the expansion of air, or elastic fluids. The volume at 32° is taken 1000, and at 212°, 1376 according to Gay Lussac's and my own experiments. As for the expansion at intermediate degrees, General Roi makes the temperature at midway of total expansion, 116°½ old scale; from the results of my former experiments, (Manch. Mem. Vol. 5, Part 2, page 599) the temperature may be estimated at 119°½; but I had not then an opportunity of having air at 32°. By more recent experiments I am convinced that dry air of 32° will expand the same quantity from that to 117° or 118° of common scale, as from the last term to 212°. According to the theory in the above Table it appears, that air of 117° will be 1188, or have acquired one half its total expansion. Now if the theory accord so well with experiment in the middle of the interval, we cannot expect it to do otherwise in the intermediate points.

The 8th column contains the force of aqueous vapours in contact with water expressed in inches of mercury, at the respective temperatures. It constitutes a geometrical progression; the numbers opposite 32° and 212°, namely, .200 and 30.0 are derived from experiments, (ibid. page 559) and the rest are determined from theory. It is remarkable that those numbers do not differ from the table just referred to, which was the result of actual experience, so much as 2° in any part; a difference that might even exist between two thermometers of the same kind.

The 9th column exhibits the force of the vapour of sulphuric ether in contact with liquid ether; which is a geometrical progression, having a less ratio than that of water. Since writing my former Essay on the Force of Steam, I am enabled to correct one of the conclusions therein contained; the error was committed by trusting to the accuracy of the common mercurial thermometer. Experience confirmed me that the force of vapour from water of nearly 212°, varied from a change of temperature as much as vapour from ether of nearly 100°. Hence I deduced this general law, namely, "that the variation of the force of vapour from all liquids is the same for the same variation of temperature, reckoning from vapour of any given force."—But I now find that 30° of temperature in the lower part of the common scale is much more than 30° in the higher: and therefore the vapours of ether and water are not subject to the same change of force by equal increments of temperature. The truth is, vapour from water, ether and other liquids, increases in force in geometrical progression to the temperature; but the ratio is different in different fluids. Ether as manufactured in the large way, appears to be a very homogeneous liquid. I have purchased it in London, Edinburgh, Glasgow and Manchester, at very different times, of precisely the same quality in respect to its vapour; namely, such as when thrown up into a barometer would depress the mercury 15 inches at the temperature of 68°. Nor does it lose any of its effect by time; I have now a barometer with a few drops of ether on the mercury, that has continued with invaried efficacy for eight or nine years. The numbers in the column between the temperatures of 20° and 80°, are the results of repeated observations on the above ether barometer for many years; those above and below are obtained from direct experiment as far as from 0 to 212°; the low ones were found by subjecting the vacuum of the barometer to an artificial cold mixture; and the higher ones were found in the manner related in my former Essays: only the highest force has been considerably increased from what I formerly had it, in consequence of supplying the manometer with more ether; it having been found to leave little or no liquid when at the temperature of 212°; and in order to obtain the maximum effect it is indispensible to have a portion of liquid remaining in contact with the vapour.

The 10th column shews the force of vapour from alcohol, or rather common spirit of wine, determined by experiment in the same way as the vapour of water. This is not a geometrical progression, probably because the liquid is not pure and homogeneous. I suspect the elastic fluid in this case is a mixture of aqueous and alcoholic vapour.

SECTION 2.

EXPANSION BY HEAT.

One important effect of heat is the expansion of bodies of every kind. Solids are least expanded; liquids more; and elastic fluids most of all. The quantities of increase in bulk have in many instances been determined; but partly through the want of a proper thermometer, little general information has been derived from particular experiments. The force necessary to counteract the expansion has not been ascertained, except in the case of elastic fluids; but there is no doubt it is very great. The quantity and law of expansion of all permanent elastic fluids have already been given; it remains then to advert to liquid and solid bodies.

In order to understand the expansion of liquids, it is expedient to premise certain propositions:

1st. Suppose a thermometrical vessel of glass, metal, &c., were filled with any liquid up to a certain mark in the stem; and that it was known the vessel and the liquid had precisely the same expansion, bulk for bulk, with the same change of temperature; then it must be evident upon a little consideration, that whatever change of temperature took place, the liquid must remain at the same mark.

2. Suppose as before, except that both bodies expand uniformly with the temperature, but the liquid at a greater rate than the vessel: then it is evident by an increase of temperature, the liquid would appear to ascend uniformly a quantity equal to the difference of the absolute expansion of the two bodies.

3. Suppose as in the last case, but that the liquid expands at a less rate than the vessel; the liquid would then descend, and that uniformly by an increase of temperature, a quantity equal to the difference of the absolute expansions.

4. Suppose as before, only the vessel now expands uniformly, and the liquid with a velocity uniformly accelerated, commencing from

rest; in this case if temperature be added uniformly, the liquid will appear to descend with a velocity uniformly retarded to a certain point, there to be stationary, and afterwards to ascend with an uniformly accelerated velocity, of the same sort as the former.—For, as the velocity with which the liquid expands is uniformly accelerative, it must successively pass through all degrees from 0 to any assigned quantity, and must therefore in some moment be the same as that of the vessel, and therefore, for that moment, the liquid must appear stationary: previously to that time the liquid must have descended by the third proposition, and must afterwards ascend, by the 2d. but not uniformly. Let the absolute space expanded by the liquid at the moment of equal velocities be denoted by 1, then that of the vessel in the same time must be 2; because the velocity acquired by an uniformly accelerating force, is such as to move a body through twice the space in the same time. It follows then that the liquid must have sunk 1, being the excess of the expansion of the vessel above that of the liquid. Again, let another portion of temperature equal to the former be added, then the absolute expansion of the liquid will be 4, reckoned from the commencement; and the expansion of the vessel also 4: the place of the liquid will be the same as at first, and therefore it must apparently ascend 1 by the 2d portion. Let a third portion of heat equal to one of the former be added, and it will make the total expansion of the liquid 9, or give 5 additional expansion, from which deducting 2, that of the vessel, there remains 3 for the apparent expansion by the 3d portion; in like manner 5 will be due for the 4th, and 7 for the 5th, &c., being the series of odd numbers. But the aggregate of these forms a series of squares, as is well known. Hence the apparent expansion will proceed by the same law as the real, only starting from a higher temperature. If the law of expansion of the liquid be such that either the addition or abstraction of temperature, that is, either heat or cold produces expansion alike, reckoned from the point of greatest density; then the apparent expansion will still be guided by the same law as the real. For, if when the liquid is at the lowest point of the scale, we withdraw a portion of heat, it ascends to 1; or is in the circumstance of greatest density, and no expansion as at the commencement; if then we withdraw another portion, it will expand 1 by hypothesis, but the vessel will contract 2, which must make the apparent expansion of the liquid 3;

by another portion it will be 5, by another 7, &c., as before.

The truth of the above proposition may be otherwise shewn thus:

Let 1, 4, 9, 16, 25, &c., represent the absolute expansions of the liquid, and p, $2p$, $3p$, $4p$, $5p$, &c., those of the vessel by equal increments of temperature, then $1-p$, $4-2p$, $9-3p$, $16-4p$, $25-5p$, &c., will represent the apparent expansion of the liquid; the differences of these last quantities, namely $3-p$, $5-p$, $7-p$, $9-p$, &c., form a series in arithmetical progression, the common difference of which is 2. But it is demonstrated by algebraists, that the differences of a series of square numbers, whose roots are in arithmetical progression, form an arithmetical progression, and that the common difference of the terms of this progression is equal to twice the square of the difference of the roots. Hence, as $2 =$ twice the square of 1, we have the above arithmetical series $3-p$, $5-p$, &c., equal to the differences of a series of squares, the common difference of the roots of which is 1.

Now to apply these principles: solid bodies are generally allowed to expand uniformly within the common range of temperature: at all events the quantity is so small compared with the expansion of liquids, such as water, that the deviation from uniformity cannot require notice in many cases. Water being supposed to expand according to the square of the temperature from that of greatest density, we may derive the following conclusions.

Cor. 1. The laws of uniformly accelerated motion, are the same as those of the expansion of water, whether absolute or apparent, the time in one denoting the temperature in the other, and the space denoting the expansion: that is, if t = time or temperature, v = velocity, and s = space or expansion: then,

t^2, or tv, or v^2 are as s.

$\frac{1}{2}\, t\, v = s$

v is as t

$\dot{s}$ is as $2\, t\, t$

$\dot{s}$ is as t, $\dot{t}$ being supposed constant, &c.

Cor. 2. The real expansion of water from maximum density for any number of degrees of temperature, is the same as the apparent expansion from apparent greatest density in any vessel for the same number of degrees. For instance, if water in a glass vessel appears to be of greatest density, or descends lowest at 42° of common scale, and appears to expand $\frac{1}{25}$ of its first volume from thence to 212°; then it may be inferred that the real expansion of water from greatest density by 170° is $\frac{1}{25}$ of its volume; so that the absolute expansion of water is determinable this way, without knowing either at what temperature its density is greatest, or the expansion of the vessel containing it.

Cor. 3. If the expansion of any vessel can be obtained; then may the temperature at which water is of greatest density be obtained; and *vice versa.* This furnishes us with an excellent method of ascertaining both the relative and absolute expansion of all solid bodies that can be formed into vessels capable of holding water.

Cor. 4. If the apparent expansion of water from maximum density for 180° were to be equalled by a body expanding uniformly, its velocity must be equal to that of water at 90°, or mid-way.—And if any solid body be found to have the same expansion as water at 10° from max. density; then its expansion for 180° must be $\frac{1}{9}$ of that of water, &c. Because in water v is as t, &c.

By graduating several glass thermometer vessels, filling them with water, exposing them to different temperatures, and comparing results, I have found the *apparent* expansion of water in glass for every 10° of the common or old scale (as I shall henceforward call it) and the new one, as under.

EXPANSION OF WATER.

OLD SCALE.		NEW SCALE.	
		5°	100227
12°	100236	15	100129
22	100090	25	100057
32	100022	35	100014
42	100000	45	100000
52	100021	55	100014
62	100083	65	100057
72	100180	75	100129
82	100312	85	100227
92	100477	95	100359
102	100672	105	100517
112	100880	115	100704
122	101116	125	100919
132	101367	135	101163
142	101638	145	101436
152	101934	155	101738
162	102245	165	102068
172	102575	175	102426
182	102916	185	102814
192	103265	195	103231
202	103634	205	103676
212	104012	215	104150
		225	104658

The whole expansion of water for 180° of temperature, reckoned from the point of greatest density, appears from the 2d Table to be $\frac{1}{21.5}$, or $21\frac{1}{2}$ parts become $22\frac{1}{2}$.

In the Edinburgh Philosophical Transactions

for 1804, Dr. Hope has given a paper on the contraction of water by heat in low temperatures. (See also Nicholson's Journal, Vol. 12.) In this paper we find an excellent history of facts and opinions relative to this remarkable question in physics, with original experiments. There appear to have been two opinions respecting the temperature at which water obtains its maximum density; the one stating it to be at the freezing point, or 32°; the other at 40°. Previously to the publication of the above essay, I had embraced the opinion that the point was 32°, chiefly from some experiments about to be related. Dr. Hope argued from his own experiments in favour of the other opinion. My attention was again turned to the subject, and upon re-examination of facts, I found them all to concur in giving the point of greatest density at the temperature 36°, or mid-way between the points formerly supposed. In two letters inserted in Nicholson's Journal, Vol. 13 and 14, I endeavoured to shew that Dr. Hope's experiments supported this conclusion and no other. I shall now shew that my own experiments on the apparent expansion of water in different vessels, coincide with them in establishing the same conclusion.

The results of my experiments, without

those deductions, were published in Nicholson's Journal, Vol. 10. Since then some small additions and corrections have been made. It may be observed that small vessels, capable of holding one or two ounces of water, were made of the different materials, and such as that glass tubes could be cemented into them when full of water, so as to resemble and act as a common thermometer. The observations follow:

		Water stationary.	Corresponding points of expansion.
1	Brown earthen ware.........	at 38°	at 32° & 44°
2	Common white ware, and stone ware,	40	32 & 48+
3	Flint glass.....................	42	32 & 52½
4	Iron...............................	42+	32 & 53—
5	Copper	45+	32 & 59
6	Brass............................	45½	32 & 60—
7	Pewter	46	32 & 60½
8	Zinc.............................	48	32 & 64+
9	Lead	49	32 & 67

As the expansion of earthen ware by heat has never before been ascertained, we cannot make use of the first and second experiments to find the temperature of greatest density; all that we can learn from them is, that the point must be below 38°.

According to Smeaton, glass expands $\frac{1}{1200}$ in length for 180° of temperature; consequently it expands $\frac{1}{400}$ in bulk. But water expands

$\frac{1}{2,1,3}$ or rather more than 18 times as much; therefore the mean velocity of the expansion of water (which is that at 90°, or half way) is 18 times more than that of glass, which is equal to the expansion of water at 42°; this last must therefore be $\frac{1}{18}$ of the former; consequently water of 42° has passed through $\frac{1}{18}$ of the temperature to the mean, or $\frac{1}{18}$ of 90° = 5°, of new scale = 4° of old scale, above the temperature at which it is absolutely of greatest density. This conclusion however cannot be accurate; for, it appears from the preceding paragraph that the temperature must be below 38°. The inaccuracy arises, I have no doubt, from the expansion of glass having been under-rated by Smeaton; not from any mistake of his, but from the peculiar nature of glass. Rods and tubes of glass are seldom if ever properly annealed; hence they are in a state of violent energy, and often break spontaneously or with a slight scratch of a file: tubes have been found to expand more than rods, and it might be expected that thin bulbs should expand more still, because they do not require annealing; hence too the great strength of thin glass, its being less brittle, and more susceptible of sudden transitions of temperature. From the above experiments it seems that the expansion due to glass, such as the bulbs of ordinary thermometers, is very little less than that of iron.

Iron expands nearly $\frac{1}{800}$ in length by 180° of heat, or $\frac{1}{265}$ in bulk; this is nearly $\frac{1}{12}$ of the expansion of water; hence 90 ÷ 12 = 7°½ of true mean temperature = 6° of common scale; this taken from 42°+, leaves 36° of common scale for the temperature at which water is of greatest density.

Copper is to iron as 3 : 2 in expansion; therefore if 6° be the allowance for iron, that for copper must be 9°; hence 45°— 9° = 36°, for the temperature as before.

Brass expands about $\frac{1}{20}$ more than copper; hence we shall have 45°½ — 9°½ = 36°, for the temperature as above.

Fine pewter is to iron as 11 : 6 in expansion, according to Smeaton; hence 46° — 11° = 35°, for the temperature as derived from the vessel of pewter: but this being a mixed metal, it is not so much to be relied upon.

Zinc expands $\frac{1}{113}$ in bulk for 180°, if we may credit Smeaton: hence water expands 5¼ times as much as zinc; and 90 ÷ 5¼ = 17° of new scale = 13°½ of old scale; whence 48° — 13°½ = 34°½ for the temperature derived from zinc. It seems highly probable that in this case the expansion of the vessel is over-rated; it was found to be less than that of lead, whereas Smeaton makes it more. The vessel was made of the patent malleable zinc of Hodson and Sylvester. Perhaps it contains a portion of tin, which will account for the deviation.

Lead expands $\frac{1}{110}$ of its bulk for 180°; water therefore expands about 5½ times as much; this gives 90 ÷ 5½ = 16°½ of new scale = 13° of old scale; whence 49° — 13° = 36°, as before.

From these experiments it seems demonstrated, that the greatest density of water is at or near the 36° of the old scale, and 37° or 38° of the new scale: and further, that the expansion of thin glass is nearly the same as that of iron, whilst that of stone ware is $\frac{2}{3}$, and brown earthen ware $\frac{1}{3}$ of the same.

The apparent expansion of mercury in a thermometrical glass for 180° I find to be .0163 from 1. That of thin glass may be stated at .0037 = $\frac{1}{270}$, which is rather less than iron, $\frac{1}{265}$. Consequently the real expansion of mercury from 32° to 212° is equal to the sum of these = 02 or $\frac{1}{50}$. De Luc makes it, .01836, and most other authors make it less; because they have all under-rated the expansion of glass. Hence we derive this proportion, .0163 : 180° :: .0037 : 41° nearly, which expresses the effect of the expansion of glass on the mercurial thermometer: that is, the mercury would rise 41° higher on the scale at the temperature of boiling water, if the glass had no expansion.—De Luc makes the expansion of a glass tube from 32° to 212° = .00083 in length, and from 32° to 122° only .00035. This inequality arises in part at least, I apprehend, from the want of equilibrium in the original fixation of glass tubes, the outside being hard when the inside is soft.

Liquids may be denominated pure when they are not decomposed by heat and cold. Solutions of salts in water cannot be deemed such; because their constitution is affected by temperature. Thus, if a solution of sulphate of soda in water be cooled, a portion of the salt crystallizes, and leaves the remaining liquid less saline than before; whereas water and mercury, when partially congealed, leave the remaining liquid of the same quality as before. Most acid liquids are similar to saline solutions in this respect. Alcohol as we commonly have it, is a solution of pure alcohol in a greater or less portion of water: and probably would be affected by congelation like other solutions. Ether is one of the purest liquids, except water and mercury. Oils, both fixed and volatile, are probably for the most part impure, in the sense we use it. Notwithstanding these observations, it is remarkable how nearly those liquids approximate to the law of expansion observed in water and mercury. Few authors have made experiments on these subjects; and their results in several instances are incorrect. My own investigations have been chiefly directed to water and mercury; but it may be proper to give the results of my enquiries on the other liquids as far as they have been prosecuted.

Alcohol expands about $\frac{1}{9}$ of its bulk for 180°, from — 8 to 172°. The relative expansions of this liquid are given by De Luc from 32° to 212°; but the results of my experiments do not seem to accord with his. According to him alcohol expands 35 parts for the first 90°, and 45 parts for the second 90°. The strength of his alcohol was such as to fire gun-powder: but this is an indefinite test. From my experiments I judge it must have been very weak. I find 1000 parts of alcohol of .817 sp. gravity at the temperature 50° became 1079 at the temperature 170° of the common mercurial scale: at 110° the alcohol is at 1039, or half a division below the true mean. When the sp. gravity is .86, I find 1000 parts at 50° become 1072 at 170°; at 110° the bulk is 1035+, whence the disproportion of the two parts of the scale is not so much in this case as 35 to 37. When the sp. gravity is ,937, I find 1000 parts become 1062 at 170°, and 1029½ at 110°; hence the ratio of the expansion becomes 29½ to 32½. When the sp. gravity is ,967, answering to 75 per cent. water, I find 1000 parts at 50° become 1040 at 170°, and 1017½ at 110°, giving a ratio of 35 to 45; which is the same as De Luc gives for alcohol. It is true he takes an interval of temperature = 180°, and I take one for 120° only; but still it is impossible to reconcile our results. As the expansion of alcohol from 172° to 212° must have been conjectural, perhaps he has over-rated it. In reporting these results I have not taken into account the expansion of the glass vessel, a large thermometrical bulb, containing about 750 grains of water, and having a tube proportionally wide; consequently the real expansions must be considered as more rather than less than above stated. The graduation of the vessel having been repeatedly examined, and being the same that was used in determining the expansion of water, I can place confidence in the results. Particular care was taken in these experiments to have the bulb and stem both immersed in water of the proposed temperature.

As alcohol of 817 sp gravity contains at

least 8 per cent. water, it is fair to infer from the above that a thermometer of pure alcohol would in no apparent degree differ from one of mercury in the interval of temperature from 50° to 170°. But when we consider that the relative expansions of glass, mercury and alcohol for this interval, are as 1, 5½ and 22 respectively, it must be obvious that the inequality of the expansion of glass in the higher and lower parts of the scale, which tends to equalise the apparent expansion of mercury, has little influence on alcohol, by reason of its comparative insignificance. Hence it may be presumed that a spirit thermometer would be more equable in its divisions than a mercurial one, in a vessel of uniform expansion. This it ought to be by theory, because the point of greatest density or congelation of alcohol is below that of mercury.

Water being densest at 36°, and alcohol at a very remote temperature below, it was to be expected that mixtures of these would be densest at intermediate temperatures, and those higher as the water prevailed; thus we find the disproportion, so observable in the expansion of water, growing greater and greater in the mixtures as they approach to pure water.

Water saturated with common salt expands as follows: 1000 parts at 32° become 1050 at 212°; at 122° it is nearly 1023, which gives the ratio of 23 to 27 for the corresponding equal intervals of mercury. This is nearly the same as De Luc's ratio of 36.3 to 43.7. This solution is said to congeal at —7°, and probably expands nearly as the square of the temperature from that point. It differs from most other saline solutions in regard to its expansion by temperature.

Olive and linseed oils expand about 8 per cent. by 180° of temperature; De Luc finds the expansion of olive oil nearly correspond to mercury; with me it is more disproportionate, nearly agreeing with water saturated with salt.

Oil of turpentine expands about 7 per cent. for 180°; it expands much more in the higher than in the lower part of the scale, as it ought to do, the freezing point being stated at 14 or 16°. The ratio is somewhere about 3 to 5. Several authors have it that oil of turpentine boils at 560°; I do not know how the mistake originated; but it boils below 212°, like the rest of the essential oils.

Sulphuric acid, sp. gravity 1.85 expands about 6 per cent. from 32° to 212°. It accords with mercury as nearly as possible in every part of the scale. Dr. Thomson says the freezing point of acid of this strength is at

—36° or below; whence it accords with the same law as water and mercury. I find that even the glacial sulphuric acid, or that of 1.78 sp. gravity, which remains congealed at 45°, expands uniformly, or nearly like the other, whilst it continues liquid.

Nitric acid, sp. gravity 1.40, expands about 11 per cent. from 32° to 212°; the expansion is nearly of the same rate as that of mercury, the disproportion not being more than 27 to 28 or thereabouts. The freezing point of acid of this strength is near the freezing point of mercury.

Muriatic acid, sp. gravity 1.137, expands about 6 per cent. from 32° to 212°; it is more disproportionate than nitric acid, as might be expected, being so largely diluted with water. The ratio is nearly 6 to 7.

Sulphuric ether expands after the rate of 7 per cent. for 180° of temperature. I have only compared the expansion of this liquid with that of mercury from 60° to 90°. In this interval it accords so nearly with mercury that I could perceive no sensible difference in their rates. It is said to freeze at —46°.

From what has been observed it may be seen that water expands less than most other liquids; yet it ought to be considered as having in reality the greatest rate of expansion.

Alcohol and nitric acid, which appear to expand so much, do not excel, or even equal water, if we estimate their expansion from the temperature of greatest density, and compare them with water in like circumstances. It is because we begin with them at 100 or 200° above the point of greatest density, and observe their expansion for 180° further, that they appear to expand so largely. Water, if it continued liquid, would expand three times as much in the *second* interval of 180° as it does in the first, reckoning from 36°.

EXPANSION OF SOLIDS.

No general law has hitherto been discovered respecting the expansion of solid bodies; but as elastic fluids and liquids appear to be subject to their respective laws in this particular, we may confidently expect that solids will be found so too. As it may be presumed that solids undergo no change of form, by the abstraction of heat, it is probable that whatever the law may be, it will respect the point at which temperature commences, or what may be called, absolute cold. It is not our present business to enquire how low this point is; but it may be observed that every phenomenon indicates it to be very low, or much lower than is commonly apprehended. Perhaps it may hereafter be demonstrated that the interval of temperature from 32° to 212° of Fahrenheit, constitutes the 10th, 15th, or 20th interval from absolute cold. Judging from analogy, we may conjecture that the expansion of solids is progressively increasing with the temperature; but whether it is a geometrical progression as elastic fluids, or one increasing as the square of the temperature, like liquids, or as the 3d or any power of the temperature, still if it be estimated from absolute cold, it must appear to be nearly uniform, or in arithmetical progression to the temperature, for so small and remote an interval of temperature as that between freezing and boiling water. The truth of this observation will appear from the following calculation: let us suppose the interval in question to be the 15th; then the real tem erature of freezing water will be 2520°, the mid-way to boiling 2610°, and boiling water 2700°, reckoned from absolute cold.

	Dif.		Dif.
$\overline{14}\rvert^2 = 196$		$\overline{14}\rvert^3 = 2744$	
	$14\frac{1}{4}$		$304\frac{5}{8}$
$\overline{14\frac{1}{2}}\rvert^2 = 210\frac{1}{4}$		$\overline{14\frac{1}{2}}\rvert^3 = 3048\frac{5}{8}$	
	$14\frac{3}{4}$		$326\frac{3}{8}$
$\overline{15}\rvert^2 = 225$		$\overline{15}\rvert^3 = 3375$	

Now the differences above represent the ratios of expansion for 90° of temperature; they are in the former case as 57 to 59, and in the latter as 14 to 15 nearly. But the temperature being supposed to be measured by the new scale, the mean is about 110° of the old scale; therefore the expansion of solids should be as 57 or 14 from 32° to 110°, and as 59 or 15 from 110° to 212° of the old scale. If these conjectures be right, the expansion of solids ought to be something greater in the lower part of the old scale, and something less in the higher part. Experience at present does not enable us to decide the question. For all practical purposes we may adopt the notion of the equable expansion of solids. Only glass has been found to expand increasingly with the temperature, and this arises probably from its peculiar constitution, as has been already observed.

Various pyrometers, or instruments for measuring the expansion of solids, have been invented, of which accounts may be seen in books of natural philosophy. Their object is to ascertain the expansion in length of any proposed subject. The longitudinal expansion being found, that of the bulk may be derived from it, and will be three times as much. Thus, if a bar of 1000 expand to 1001 by a certain temperature; then 1000 cubic inches of the same will become 1003 by the same temperature.

The following Table exhibits the expansion of the principal subjects hitherto determined, for 180° of temperature; that is, from 32° to 212° of Fahrenheit. The bulk and length of the articles at 32° are denoted by 1.

	EXPANSION.	
SOLIDS.	In bulk.	In length.
Brown earthen ware	.0012 = $\frac{1}{800}$	$\frac{1}{2400}$
Stone ware	.0025 = $\frac{1}{400}$	$\frac{1}{1200}$
Glass—rods and tubes	.0025 = $\frac{1}{400}$	$\frac{1}{1200}$ †
—— bulbs (thin)	.0037 = $\frac{1}{270}$	$\frac{1}{810}$
Platinum	.0026 = $\frac{1}{385}$	$\frac{1}{1155}$ ‡
Steel	.0034 = $\frac{1}{294}$	$\frac{1}{882}$ †
Iron	.0038 = $\frac{1}{263}$	$\frac{1}{790}$ †
Gold	.0042 = $\frac{1}{238}$	$\frac{1}{714}$ *
Bismuth	.0042 = $\frac{1}{238}$	$\frac{1}{714}$ †
Copper	0051 = $\frac{1}{196}$	$\frac{1}{588}$ †
Brass	.0056 = $\frac{1}{178}$	$\frac{1}{533}$ †
Silver	.0060 = $\frac{1}{160}$	$\frac{1}{480}$ *
Fine Pewter	.0068 = $\frac{1}{140}$	$\frac{1}{420}$ †
Tin	.0074 = $\frac{1}{133}$	$\frac{1}{400}$ †
Lead	.0086 = $\frac{1}{115}$	$\frac{1}{348}$ †
Zinc	.0093 = $\frac{1}{108}$	$\frac{1}{322}$ †
LIQUIDS.		
Mercury	.0200 = $\frac{1}{50}$	
Water	.0466 = $\frac{1}{21.5}$	
Water sat. with salt	.0500 = $\frac{1}{20}$	
Sulphuric acid	.0600 = $\frac{1}{17}$	
Muriatic Acid	.0600 = $\frac{1}{17}$	
Oil of turpentine	.0700 = $\frac{1}{14}$	
Ether	.0700 = $\frac{1}{14}$	
Fixed oils	.0800 = $\frac{1}{12.5}$	
Alcohol	.0110 = $\frac{1}{9}$	
Nitric acid	0110 = $\frac{1}{9}$	
ELASTIC FLUIDS.		
Gases of all kinds	.376 = $\frac{3}{8}$	

† Smeaton. * Ellicott. ‡ Borda.

Wedgwood's Thermometer.

The spirit thermometer serves to measure the greatest degrees of cold we are acquainted with, and the mercurial thermometer measures 400° above boiling water, by the old scale, or about 250° by the new one, at which temperature the mercury boils. This is short of red heat, and very far short of the highest attainable temperature. An instrument to measure high temperatures is very desirable; and Mr. Wedgwood's is the best we have yet; but there is still great room for improvement. Small cylindrical pieces of clay, composed in the manner of earthen ware, and slightly baked, are the thermometrical pieces. When used, one of them is exposed in a crucible to the heat proposed to be measured, and after cooling, it is found to be contracted, in proportion to the heat previously sustained; the quantity of contraction being measured, indicates the temperature. The whole range of this thermometer is divided into 240 equal degrees, each of which is calculated to be equal to 130° of Fahrenheit. The lowest, or 0, is found about 1077° of Fahrenheit (supposing the common scale continued above boiling mercury,) and the highest 32277°. According to the new views of temperature in the preceding pages, there is reason to think these numbers are much too large.

The following Table exhibits some of the more remarkable temperatures in the whole range, according to the present state of our knowledge.

	Wedg.
Extremity of Wedgwood's thermometer..............	240°
Pig iron, cobalt and nickel, melt from 130° to......	150
Greatest heat of a Smith's forge......................	125
Furnaces for glass and earthen ware, from 40 to......	124
Gold melts	32
Settling heat of flint glass............................	29
Silver melts...	28
Copper melts	27
Brass melts	21
Diamond burns	14
Red heat visible in day-light.........................	0
	Fahrenheit. old scale.
Hydrogen and charcoal burn 800° to	1000°
Antimony melts...	809
Zinc	700
Lead	612
Mercury boils	600
Linseed oil boils..	600
Sulphuric acid boils....................................	590
Bismuth...	476
Tin	442
Sulphur burns slowly..................................	303
Nitric acid boils	240
Water and essential oils boil	212
Bismuth 5 parts, tin 3 and lead 2, melt...............	210

	Fahrenheit.
Alcohol boils	174°
Bees wax melts	142
Ether boils	98
Blood heat 96° to	98
Summer heat in this climate 75° to	80
Sulphuric acid (1.78) when congealed, begins to melt	45
Mixture of ice and water	32
Milk freezes	30
Vinegar freezes	28
Strong wines freeze about	20
Snow 3 parts, salt 2	—7
Cold observed on the snow at Kendal, 1791	—10
Ditto at Glasgow, 1780	—23
Mercury freezes	—39
Greatest artificial cold observed	—90

SECTION 3.

ON THE

SPECIFIC HEAT OF BODIES.

If the whole quantity of heat in a measure of water of a certain temperature be denoted by 1, that in the same measure of mercury will be denoted by .5 nearly: hence the specific heats of water and mercury, *of equal bulks*, may be signified by 1 and .5 respectively.

If the specific heats be taken from *equal weights* of the two liquids; then they will be denoted by 1 and .04 nearly; because we have to divide .5 by 13.6, the specific gravity of mercury.

That bodies differ much in their specific heats, is manifest from the following facts.

1. If a measure of mercury of 212° be mixed with a measure of water of 32°, the mixture will be far *below* the mean temperature.

2. If a measure of mercury of 32° be mixed with a measure of water of 212°, the mixture will be far *above* the mean.

3. If two equal and like vessels be filled, the one with hot water, the other with hot mercury; the latter will cool in about half the time of the former.

4. If a measure of sulphuric acid be mixed with a measure of water of the same temperature, the mixture will assume a temperature about 240° higher.

These facts clearly shew that bodies have various affinities for heat, and that those bodies which have the strongest attraction or affinity for heat, possess the most of it in like circumstances; in other words, they are said to have the greatest *capacity for heat*, or the greatest specific heat. It is found too that the same body changes its capacity for heat, or apparently assumes a new affinity, with a change of form. This no doubt arises from a new arrangement or disposition of its ultimate particles, by which their atmospheres of heat are influenced: Thus a solid body, as ice, on becoming liquid, acquires a larger capacity for heat, even though its bulk is diminished; and a liquid, as water, acquires a larger capacity for heat on being converted into an elastic fluid; this last increase is occasioned, we may conceive, solely by its being increased in bulk, in consequence of which every atom of liquid possesses a larger sphere than before.

A very important enquiry is, whether the same body in the same state undergoes any change of capacity by change of temperature. Does water, for instance, at 32° possess the same capacity for heat, as at 212°, and through all the intermediate degrees? Dr. Crawford, and most writers after him, contend, that the capacities of bodies in such circumstances are *nearly* permanent. As an outline of doctrine this may be admitted; but it is requisite, if possible, to ascertain, whether the small change of capacity induced by temperature, is such as to *increase* the capacity, or to *diminish* it; and also, whether the increase or diminution is uniform or otherwise. Till this point is settled, it is of little use to mix water of 32° and 212°, with a view to obtain the true mean temperature.

That water *increases* in its capacity for heat with the increase of temperature, I consider demonstrable from the following arguments: 1st. A measure of water of any one temperature being mixed with a measure at any other temperature, the mixture is less than two measures. Now a condensation of volume is a certain mark of diminution of capacity and increase of temperature, whether the condensation be the effect of chemical agency, as in the mixture of sulphuric acid and water, or the effect of mechanical pressure, as with elastic fluids. 2. When the same body suddenly changes its capacity by a change of form, it is always from a *less* to a *greater*, as the temperature ascends; for instance, ice, water and vapour. 3. Dr. Crawford acknowledges from his own experience, that dilute sulphuric acid, and most other liquids he tried, were found to increase in their capacity for heat with the increase of temperature.

Admitting the force of these arguments, it follows that when water of 32° and 212° are mixed, and give a temperature denoted by 119° of the common thermometer, we must conclude that the true mean temperature is somewhere *below* that degree. I have already assigned the reasons why I place the mean at 110°.

With respect to the question whether water varies uniformly or otherwise in its capacity, I am inclined to think the increase, in this respect, will be found nearly proportional to the increase in bulk, and consequently will be four times as much at 212° as at the mean. Perhaps the expressions for the bulk may serve for the capacity; if so, the ratios of the capacities at 32°, 122° and 212° of the new scale, may be denoted by 22, 22¼ and 23. I should rather expect, however, that the ratios are much nearer equality, and that 200, 201 and 204, would be nearer the truth.*

* In the Lectures I delivered in Edinburgh and Glasgow in the spring of 1807, I gave it as my opinion that the capacity of water at 32° was to that at 212°, as 5 to 6, nearly. The opinion was founded on the fact I had just before observed, that a small mercurial thermometer at the temperature 32° being plunged into boiling water, rose to 202° in 15″; but the same at 212° being plunged into ice-cold water, was 18″ in descending to 42°; estimating the capacities to be reciprocally as the times of cooling, it gave the ratio of 5 to 6. On more mature consideration I am persuaded this difference is occasioned, not so much by the difference of capacities, as by the different degrees of fluidity. Water of 212° is more fluid than water of 32°, and distributes the temperature with greater facility. By a subsequent experiment too, I find, that mercury cools a

Dr. Crawford, when investigating the accuracy of the common thermometer, was aware, that if equal portions of water of different temperatures were mixed together, and the thermometer always indicated the mean, this was not an infallible proof of its accuracy. He allows that if water have an increasing capacity, and the mercury expand increasingly with the temperature, an equation may be formed so as to deceive us. This is in fact the case in some degree; and he appears to have been deceived by it. Yet the increased capacity of water, is by no means sufficient to balance the increased expansion of the mercury, as appears from the following experiments.

I took a vessel of tinned iron, the capacity of which was found to be equal to 2 oz. of water; into this were put 58 ounces of water, making the sum = 60 ounces of water. The whole was raised to any proposed temperature, and then two ounces of ice were put in and melted; the temperature was then observed, as follows:

thermometer twice as fast as water, though it has but half its capacity for heat; the times in which a thermometer is in cooling in fluids, are not, therefore, tests of their specific heats.

60 oz. water of 212° + 2 oz. ice of 32°, gave 200°½
60 oz. water of 130° + 2 oz. ice of 32°, gave 122°
60 oz. water of 50° + 2 oz. ice of 32°, gave 45°.3

From the first of these, 30 parts of water lost 11°½ each, or 345°, and 1 part water of 32° gained 168°½; the difference 345 — 168½ = 176°½, expresses the number of degrees of temperature (such as are found between 200 and 212 of the old scale) entering into ice of 32° to convert it into water of 32°. Similar calculations being made for the other two, we find in the second, 150°, and in the third, 128°. These three resulting numbers are nearly as 5, 6 and 7. Hence it follows that as much heat is necessary to raise water 5° in the lower part of the old scale, as is required to raise it 7° in the higher, and 6° in the middle.*

Methods of finding the Specific Heats of Bodies.

The most obvious method of ascertaining the specific heats of bodies that have no chemical

* Perhaps the above results may account for the diversity in authors respecting the quantity of *latent* heat (improperly so called) in water. Respecting the doctrine of Black on Latent Heat, see an excellent note of Leslie. (Inquiry, page 529.)

affinity for water, is to mix equal weights of water, and any proposed body of two known temperatures, and to mark the temperature of the mixture. Thus, if a pound of water of 32°, and a pound of mercury of 212°, be mixed, and brought to a common temperature, the water will be raised *m* degrees, and the mercury depressed *n* degrees; and their capacities or specific heats will be inversely as those numbers; or, $n : m ::$ specific heat of water : specific heat of mercury. In this way Black, Irvine, Crawford and Wilcke, approximated to the capacities of various bodies. Such bodies as have an affinity for water, may be confined in a vessel of known capacity, and plunged into water so as to be heated or cooled, as in the former case.

The results already obtained by this method are liable to two objections: 1st. the authors presume the capacities of bodies while they retain their form are permanent; that is, the specific heat increases exactly in proportion to the temperature; and 2d, that the common mercurial thermometer is a true test of temperature. But it has been shewn that neither of these positions is warrantable.

The calorimeter of Lavoisier and Laplace was an ingenious contrivance for the purpose of investigating specific heat; it was calculated to shew the quantity of ice which any body heated to a given temperature could melt. It was therefore not liable to the 2d objection above. Unfortunately this instrument does not seem to have answered well in practice.

Meyer attempted to find the capacities of dried woods, by observing the times in which given equal volumes of them were in cooling. These times he considered as proportionate to the capacities bulk for bulk; and when the times were divided by the specific gravities, the quotient represented the capacities of equal weights. (Annal. de Chemie Tom. 30). Leslie has since recommended a similar mode for liquids, and given us the results of his trials on 5 of them. From my own experience I am inclined to adopt this method as susceptible of great precision. The times in which bodies cool in like circumstances appear to be ascertainable this way with uncommon exactness, and as they are mostly very different, a very small error is of little consequence. The results too I find to agree with those by mixture; and they have the advantage of not being affected by any error in the thermometric scale.

The formulæ for exhibiting the phenomena of the specific heats of bodies are best conceived from the contemplation of cylindrical vessels of unequal bases. (See plate 1. Fig. 1). Supposing heat to be represented by a quantity of liquid in each vessel, and temperature by the height of the liquid in the vessel, the base denoting the zero or total privation of heat; then the specific heats of bodies at any given temperature, x, will be denoted by multipling the area of the several bases by the height or temperature, x. Those specific heats too will be directly as the bases, or as the increments of heat necessary to produce equal changes of temperature.

Let w and W = the weights of two cold and hot bodies; c and C their capacities for heat at the same temperature (or the bases of the cylinders); d = the difference of the temperature of the two bodies before mixture, reckoned in degrees; m = the elevation of the colder body, and n = the depression of the warmer after mixture, (supposing them to have no chemical action); then we obtain the following equations.

$$1.\quad m + n = d.$$

$$2.\quad m = \frac{W\,C\,d}{w\,c + W\,C}.$$

$$3.\quad C = \frac{w\,c\,m}{W\,n}.$$

$$4. \quad c = \frac{W\,C\,n}{w\,m}$$

If $C = c$, then, $5. \quad m = \frac{W\,d}{W+w}.$

If $W = w$, then, $6. \quad C = \frac{c\,m}{n}.$

To find the zero, or point of absolute privation of temperature, from observations on the change of capacity in the same body. Let c = the less, and C = the greater capacity, m = the number of degrees of the less capacity requisite to produce the change in equal weights, n = the number of degrees of the greater capacity, x = the whole number of degrees of temperature down to zero; then,

$$7. \quad C\,x - c\,x = C\,n = c\,m.$$

$$8. \quad x = \frac{C\,n}{C - c} = \frac{c\,m}{C - c}.$$

To find the zero from mixing two bodies of the same temperature which act chemically, and produce a change of temperature. Let w, W, c, C & x be as before; let M = capacity of the mixture, and n = the degrees of heat or cold produced: then the quantity of heat in both bodies will be $= (c\,w + C\,W)\,x$ $= (w + W)\,M\,x \pm (w + W)\,M\,n.$

$$9. \text{ and } x = \frac{(w + W)\,M\,n}{(c\,w + C\,W) \backsim (w + W)\,M}$$

It is to be regretted that so little improvement has been made for the last fifteen years in this department of science. Some of the earliest and most incorrect results are still obtruded upon the notice of students; though with the least reflection their errors are obvious. I have made great number of experiments with a view to enlarge, but more especially, to correct the Tables of Specific Heat. It may be proper to relate some of the particulars. For liquids I used an egg-shaped thin glass vessel, capable of holding eight ounces of water; to this was adapted a cork, with a small circular hole, sufficient to admit the stem of a delicate thermometer tube, which had two small marks with a file, the one at 92°, and the other at 82°, both being above the cork; when the cork was in the neck of the bottle, the bulb of the thermometer was in the centre of the internal capacity. When an experiment was made the bottle was filled with the proposed liquid, and heated a little above 92°; it was then suspended in the middle of a room, and the time accurately noted when the thermometer was at 92°, and again when it was 82°, another thermometer at the same time indicating the temperature of the air in the room. The capacity of the glass vessel was found = $\frac{2}{3}$ oz. of water.

The mean results of several experiments were as follow:

Air in the Room 52°.

	Minutes.
Water cooled from 92° to 82°, in	29
Milk (1.026)	29
Solution of carbonate of potash (1.30)	28½
Solution of carbonate of ammonia (1.035)	28½
Ammoniacal solution (.948)	28½
Common vinegar (1.02)	27½
Solution of common salt, 88 W + 32 S. (1.197)	27
Solution of soft sugar, 6 W. + 4 S. (1.17)	26¼
Nitric acid (1.20)	26½
Nitric acid (1.30)	25½
Nitric acid (1.36)	25
Sulphuric acid (1.844) and water, equal bulks (1.535)	23½
Muriatic acid (1.153)	21
Acetic acid (1.056) from Acet. Cop.	21
Sulphuric acid (1.844)	19½
Alcohol (.85)	19½
Ditto (.817)	17½
Ether sulphuric (.76)	15½
Spermaceti oil (.87)	14

These times would express accurately the specific heats of the several bodies, bulk for bulk, provided the heat of the glass vessel did not enter into consideration. But as the heat of that was proved to be equal to $\frac{2}{3}$ of an ounce of water, or to $\frac{4}{3}$ of an ounce measure of oil, it is evident we must consider the heat disengaged in the 1st experiment, as from 8 $\frac{2}{3}$ ounces of water, and in the last as from 9 $\frac{1}{3}$ ounce measures of oil. On this account the numbers below 29 will require a small reduction, before they can be allowed to represent the times of cooling of *equal bulks* of the different liquids; in the last experiment the reduction will be one minute, and less in all the preceding ones.

It may be proper to observe, that the above results do not depend upon one trial of the several articles; most of the experiments were repeated several times, and the times of cooling were found not to differ more than half a minute; indeed, in general, there was no sensible differences. If the air in the room was, in any case, a little above or below 52°, the due allowance was made.

I found the specific heat of mercury, by mixture with water, and by the time of its cooling in a smaller vessel than the above, to be to that of water of equal bulk, as, .55 to 1 nearly.

I found the specific heats of the metals and other solids after the manner of Wilcke and

Crawford; having procured a goblet, of very thin glass and small stem, I found its capacity for heat; then put water into it, such that the water, together with the value of the glass in water, might be equal to the weight of the solid. The solid was raised to 212°, and suddenly plunged into the water, and the specific heats of equal weights of the solid and the water, were inferred to be inversely as the changes of temperature which they experienced, according to the 6th formula. Some regard was paid to the correction, on account of the error of the common thermometer, which was used on the occasion. The solids I tried were iron, copper, lead, tin, zinc, antimony, nickel, glass, pitcoal, &c. The results differed little from those of Wilcke and Crawford; their numbers may, therefore, be adopted without any material error, till greater precision can be attained. In the following Table I have not carried the decimals beyond two places; because present experience will not warrant further extension: the first place of decimals may, I believe, be relied upon as accurate, and the second generally so, but in a few instances it may, perhaps, be 1 or 2 wrong; except from this observation, the specific heats of the gases by Crawford, on which I shall further remark.

TABLE OF SPECIFIC HEATS.

GASES.	equal weights	eq. blks
Hydrogen - -	21.40*	.002
Oxygen - -	4.75*	.006
Common air - -	1.79*	.002
Carbonic acid - -	1.05*	.002
Azotic - - -	.79*	.001
Aqueous vapour - -	1.55*	.001
LIQUIDS.		
Water - - -	1.00	1.00
Arterial blood - -	1.03*	
Milk (1.026) -	.98	1.00
Carbonat. of ammon. (1:035)	.95	.98
Carbonat. of potash (1 30)	.75	.98
Solut. of ammonia (948)	1.03	.98
Common vinegar (1 02) -	.92	.94
Venous blood - -	.89*	
Solut. of common salt (1.197)	.78	.93
Solut. of sugar (1.17) -	.77	.90
Nitric acid (1.20) - -	.76	.96
Nitric acid (1.30) - -	.68	.88
Nitric acid (1.36) - -	.63	.85
Nitrate of lime (1.40) -	.62	.87
Sulph acid and water, equal b	.52	:80
Muriatic acid (1 153) -	.60	.70
Acetic acid (1.056) -	:66	.70
Sulphuric acid (1.844) -	.35	.65
Alcohol (.85) - -	.76	.65
Ditto (.817) - -	.70	.57
Sulphuric ether (76) -	.66	.50
Spermaceti oil (.87) -	.52	.45
Mercury - -	.04	.55

SOLIDS.	eq. wts.	eq. blks
Ice - -	.90 ?	.83
Dried woods, and other vegetable substances, from .45 to -	.65	
Quicklime -	.30	
Pit-coal (1 27) -	.28	.36
Charcoal - -	.26*	
Chalk - -	.27	.67
Hydrat. lime -	.25	
Flint glass (2.87) -	.19	.55
Muriate of soda -	.23	
Sulphur - -	.19	
Iron - -	.13	1.00
Brass - -	.11	.97
Copper - -	.11	.98
Nickel - -	.10	.78
Zinc - -	.10	.69
Silver -	.08	.84
Tin - -	.07	.51
Antimony - -	.06	.40
Gold - -	.05	.97
Lead - -	.04	.45
Bismuth - -	.04	.40

Oxides of the metals surpass the metals themselves, according to Crawford.

Remarks on the Table.

The articles marked * are from Crawford. Notwithstanding the ingenuity and address displayed in his experiments on the capacities of the elastic fluids, there is reason to believe his results are not very near approximations to the truth; we can never expect accuracy when it depends upon the observation of 1 or 2 tenths of a degree of temperature after a tedious and complicated process. Great merit is undoubtedly due to him for the attempt.— The difference between arterial and venous blood, on which he has founded the beautiful system of animal heat, is remarkable, and deserves further enquiry.

From the observed capacities of water, solution of ammonia, and the combustibles, into which hydrogen enters, together with its small specific gravity, we cannot doubt but that this element possesses a very superior specific heat. Oxygen, and azote likewise, undoubtedly stand high, as water and ammonia indicate; but the compound of these two elements denominated nitric acid, being so low, compared with the same joined to hydrogen, or water and ammonia, we must conclude that the superiority of the two last articles is chiefly due to the hydrogen they contain. The elements, charcoal and sulphur, are remarkably low, and carry their character along with them into compounds, as oil, sulphuric acid, &c.

Water appears to possess the greatest capacity for heat of any pure liquid yet known, whether it be compared with equal bulks or weights; indeed it may be doubted, whether any solid or liquid whatever contains more heat than an equal bulk of water of the same temperature. The great capacity of water arises from the strong affinity, which both its elements, hydrogen, and oxygen, have for heat. Hence it is that solutions of salts in water, contain generally less heat in a given volume than pure water: for, salts increase the volume of water as well as the density, and having mostly a small capacity for heat, they enlarge the volume of the water more than proportional to the heat they contribute.

Pure ammonia seems to possess a high specific heat, judging from the aqueous solution, which contains only about 10 per cent.—If it could be exhibited pure in a liquid form, it would probably exceed water in this particular.

The compounds of hydrogen and carbon, under the characters of oil, ether and alcohol, and the woods, all fall below the two last mentioned; the reason seems to be, because charcoal is an element of a low specific heat.

The acids form an interesting class of bodies in regard to their specific heats.—Lavoisier is the only one who is nearly correct in regard to nitric acid; he finds the specific heat of the acid 1.3 to be .66; this with some other of his results I find rather too low. It is remarkable that the water in acid of this strength is 63 per cent. and should have nearly as much heat in it as the compound is found to have, whence it should seem that the acid loses the principal part of its heat on combining with water. This is still more observable in muriatic acid, which contains 30 per cent. of water, and its specific heat is only .60; whence not only the heat of the acid gas, but part of that in the water is expelled on the union; this accounts for the great heat produced by the union of this acid gas with water.

The specific heat of sulphuric acid has been well approximated by several.—Gadolin and Leslie make it .34, Lavoisier .33+; Crawford finds it .43, but he must probably have had a diluted acid.

Common vinegar, being water with 4 or 5 per cent. of acid, does not differ materially from water in its specific heat; it has been stated at .39 and at .10; but such results do not require animadversion. The acetic acid

I used contained 33 per cent. pure acid, this acid therefore, in combining with water, expels much heat.

Quicklime is determined by Lavoisier and Crawford to be .22; I think they have underrated it: I find quicklime to impart as much or more heat than carbonate of lime, when inclosed in a vessel and plunged in water, or when mixed with oil. Hydrat of lime (that is, quicklime 3 parts and water 1 part, or dry slaked lime) is fixed at .28 by Gadolin: it was .25 by my first experiments; but I since find I have underrated it. The subject will be adverted to in a future section.

SECTION 4.

THEORY OF THE SPECIFIC HEAT OF ELASTIC FLUIDS.

Since the preceding section was printed off, I have spent some time in considering the constitution of elastic fluids with regard to heat. The results already obtained cannot be relied upon; yet it is difficult to conceive and execute experiments less exceptionable than those of Crawford. It is extremely important, however, to obtain the exact specific heat of elastic fluids, because the phenomena of combustion and of heat in general, and consequently a great part of chemical agency, are intimately connected therewith.

In speaking of the uncertainty of Crawford's results on the specific heat of elastic fluids, it must not be understood that *all* of them are equally implicated. The reiterated experiments on the heat given out by the combustion of hydrogen, in which it was found that 11 measures of mixed gases, when fired by electricity heated 20.5 measures of water 2°.4 (page 263) at a medium, were susceptible of very considerable accuracy, and are therefore entitled to credit. The comparative heat of atmospheric air and water, which rested on the observance of nearly $\frac{1}{4}$ of a degree of temperature, is probably not very far from the truth; but the very small differences in the heats communicated by equal bulks of oxygen, hydrogen, carbonic acid, azotic gas and common air, together with the great importance of those differences in the calculation, render the results very uncertain. He justly observes, that if we suppose the heats imparted by equal bulks of these gases to be equal, it will not affect his doctrine. The tenor of it necessarily led him to estimate the heat of oxygen high, compared with equal weights of carbonic acid and aqueous vapour, and of azotic gas or *phlogisticated* air, as it was then called, under the idea of its being an opposite to oxygen or *dephlogisticated* air. Indeed his deductions respecting azotic gas, are not consistent with his experiments: for he makes no use of experiments 12 and 13, which are the only direct ones for the purpose, but he infers the heat of azotic gas from the observed difference between oxygen and common air. The result gives it less than half that of common air; whereas from the 13th experiment, scarcely any sensible difference was perceived between them. He has in all probability much underrated it; but his errors in this respect whatever they may be, do not affect his system.

When we consider that all elastic fluids are equally expanded by temperature, and that liquids and solids are not so, it should seem that a general law for the affection of elastic fluids for heat, ought to be more easily deducible and more simple than one for liquids, or solids.—There are three suppositions in regard to elastic fluids which merit discussion.

1. *Equal weights of elastic fluids may have the same quantity of heat under like circumstances of temperature and pressure.*

The truth of this supposition is disproved

by several facts: oxygen and hydrogen upon their union give out much heat, though they form steam, on elastic fluid of the same weight as the elements composing it. Nitrous gas and oxygen unite under similar circumstances. Carbonic acid is formed by the union of charcoal, a substance of low specific heat, with oxygen; much heat is given out, which must be principally derived from the oxygen; if then the charcoal contain little heat, and the oxygen combining with it be reduced, the carbonic acid must be far inferior in heat to an equal weight of oxygenous gas.

2. *Equal bulks of elastic fluids may have the same quantity of heat with the same pressure and temperature.*

This appears much more plausible; the diminution of volume when a mixture of oxygen and hydrogen is converted into steam, may be occasioned by a proportionate diminution of the absolute heat; the same may be said of a mixture of nitrous gas and oxygen. The minute differences observed by Crawford, may have been inaccuracies occasioned by the complexity of his experiments.—But there are other considerations which render this supposition extremely improbable, if they do not altogether disprove it. Carbonic acid contains its own bulk of oxygen; the heat given out at its formation must therefore be exactly equal to the whole heat previously contained in the charcoal on this supposition; but the heat by the combustion of one pound of charcoal seems, at least, equal to the heat by the combustion of a quantity of hydrogen sufficient to produce one pound of water, and this last is equal to, or more than the heat retained by the water, because steam is nearly twice the density of the elastic mixture from which it is produced; it should therefore follow, that charcoal should be found of the same specific heat as water, whereas it is only about $\frac{1}{4}$ of it. Were this supposition true, the specific heats of elastic fluids of equal weights would be inversely as their specific gravities.—If that of steam or aqueous vapour were represented by 1, oxygen would be .64, hydrogen 8.4, azote .72, and carbonic acid .46.—But the supposition is untenable.

3. *The quantity of heat belonging to the ultimate particles of all elastic fluids, must be the same under the same pressure and temperature.*

It is evident the number of ultimate particles or molecules in a given weight or volume of one gas is not the same as in another: for, if equal measures of azotic and oxygenous gases were mixed, and could be instantly united chemically, they would form nearly two measures of nitrous gas, having the same weight as the two original measures; but the number of ultimate particles could at most be one half of that before the union. No two elastic fluids, probably, therefore, have the same number of particles, either in the same volume or the same weight. Suppose, then, a given volume of any elastic fluid to be constituted of particles, each surrounded with an atmosphere of heat repelling each other through the medium of those atmospheres, and in a state of equilibrium under the pressure of a constant force, such as the earth's atmosphere, also at the temperature of the surrounding bodies; suppose further, that by some sudden change each malecule of air was endued with a stronger affinity for heat; query the change that would take place in consequence of this last supposition? The only answer that can be given, as it appears to me, is this.—The particles will condense their respective atmospheres of heat, by which their mutual repulsion will be diminished, and the external pressure will therefore effect a proportionate condensation in the volume of air: neither an increase nor diminution in the quantity of heat around each malecule, or around the whole, will take place. Hence the truth of the supposition, or as it may now be called, proposition, is demonstrated.

Corol. 1. The specific heats of equal *weights* of any two elastic fluids, are inversely as the weights of their atoms or molecules.

2. The specific heats of equal *bulks* of elastic fluids, are directly as their specific gravities, and inversely as the weights of their atoms.

3. Those elastic fluids that have their atoms the most condensed, have the strongest attraction for heat; the greater attraction is spent in accumulating more heat in a given space or volume, but does not increase the quantity around any single atom.

4. When two elastic atoms unite by chemical affinity to form one elastic atom, one half of their heat is disengaged. When three unite, then two thirds of their heat is disengaged, &c. And in general, when m elastic particles by chemical union become n; the heat given out is to the heat retained as $m—n$ is to n.

One objection to this proposition it may be proper to obviate: it will be said, an increase in the specific attraction of each atom must produce the same effect on the system as an increase of external pressure. Now this last is known to express or give out a quantity of the absolute heat; therefore the former must do the same. This conclusion must be admitted; and it tends to establish the truth of the preceding proposition. The heat expressed by doubling the density of any elastic fluid amounts to about 50°, according to my former experiments; this heat is not so much as one hundreth part of the whole, as will be shewn hereafter, and therefore does not materially affect the specific heat: it seems to be merely the interstitial heat amongst the small globular molecules of air, and scarcely can be said to belong to them, because it is equally found in a vacuum or space devoid of air, as is proved by the increase of temperature upon admitting air into a vacuum.

Before we can apply this doctrine to find the specific heat of elastic fluids, we must first ascertain the relative weights of their ultimate particles. Assuming at present what will be proved hereafter, that if the weight of an atom of hydrogen be 1, that of oxygen will be 7, azote 5, nitrous gas 12, nitrous oxide 17, carbonic acid 19, ammoniacal gas 6, carburetted hydrogen 7, olefiant gas 6, nitric acid 19, carbonic oxide 12, sulphuretted hydrogen 16, muriatic acid 22, aqueous vapour 8, ethereal vapour 11, and alcoholic vapour 16; we shall have the specific heats of the several elastic fluids as in the following table. In order to compare them with that of water, we shall further assume the specific heat of water to that of steam as 6 to 7, or as 1 to 1.166.

Table of the specific heats of elastic fluids.

Hydrogen	9.382	Olefiant gas	1.555
Azote	1.866	Nitric acid	.491
Oxygen	1.333	Carbonic oxide	.777
Atmos. air	1.759	Sulph. hydrogen	.583
Nitrous gas	.777	Muriatic acid	.424
Nitrous oxide	549	Aqueous vapour	1.166
Carbonic acid	.491	Ether. vapour	.848
Ammon. gas	1.555	Alcohol. vapour	.586
Carb. hydrogen	1.333	Water	1.000

Let us now see how far these results will accord with experience. It is remarkable that the heat of common air comes out nearly the same as Crawford found it by experiment; also, hydrogen excels all the rest as he determined; but oxygen is much lower and azote higher. The principles of Crawford's doctrine of animal heat and combustion, however, are not at all affected with the change. Besides the reason already assigned for thinking that azote has been rated too low, we see from the Table, page 62, that ammonia, a compound of hydrogen and azote, has a higher specific heat than water, a similar compound of hydrogen and oxygen.

Upon the whole, there is not any established fact in regard to the specific heats of bodies, whether elastic or liquid, that is repugnant to the above table as far as I know; and it is to be hoped, that some principle analogous to the one here adopted, may soon be extended to solid and liquid bodies in general.

SECTION 5.

ON THE QUANTITY OF HEAT EVOLVED BY COMBUSTION.

When certain bodies unite chemically with oxygen, the process is denominated *combustion*, and is generally accompanied with the evolution of heat, in consequence of the diminished capacities of the products. The fine attempt of Lavoisier and Laplace to find the quantities of heat disengaged during different species of combustion, has not been followed up with the attention it deserves. Perhaps this may have been owing to the supposed necessity of using the calorimeter of the above philosophers, and to a notion that its results are not always to be depended upon. Much important information may, however, be obtained on this subject by the use of a very simple apparatus, as will appear from what follows:

I took a bladder, the bulk of which, when extended with air, was equal to 30000 grains of water; this was filled with any combustible gas, and a pipe and stop-cock adapted to it: a tinned vessel, capable of containing 30000 grains of water was provided, and its capacity for heat being found, so much water was put into it as to make the vessel and water together, equal to 30000 grains of water. The gas was lighted, and the point of the small flame was applied to the concavity of the bottom of the tinned vessel, till the whole of the gas was consumed; the increase of the temperature of the water was then carefully noted; whence the effect of the combustion of a given volume of gas, of the common pressure and temperature, in raising the temperature of an equal volume of water, was ascertained, except a very small loss of heat by radiation, &c. which this method must be liable to, and which probably does not exceed $\frac{1}{8}$ or $\frac{1}{10}$th of the whole.

The mean results of several trials of the different gases are stated below; when the experiments are performed with due care, there is scarcely any sensible differences in the results with the same species of gas. The point of the flame should just touch the bottom of the vessel.

Hydrogen, combustion of it raises an
equal volume of water ....4°.5
Coal gas, or carburetted hydrogen........10.
Olefiant gas ....14.
Carbonic oxide 4.5

Oil, alcohol, and ether, were burned in a lamp, &c. and the effect observed as under:

Oil, spermaceti, combustion of 10 grs.
raised 30000 grs. water ....5°.
— of turpentine (much smoke unburnt) 3
Alcohol (.817) 2.9
Ether, sulphuric 3.1
Tallow and wax...............................5.2

Phosphor. —10grs. heated 30000grs. water 3
Charcoal 2
Sulphur.. 1
Camphor....................................... 3.5
Caoutchouc......................................2.1

The five last articles were placed upon a convenient stand, and burned under the vessel of water; except charcoal, a piece of which was ignited, then weighed, and the combustion was maintained by a gentle blast from a blow-pipe, directing the heat as much as possible upon the bottom of the vessel; after the operation it was again weighed, and the loss ascertained; the result never amounted to 2° for ten grains, but generally approached it nearly.

In order to exhibit the comparative effects more clearly, it may be proper to reduce the articles to a common weight, and to place along with them the quantity of oxygen known to combine with them. The quantity of heat given out may well be expressed by the number of pounds of ice which it would melt, taking it for granted that the quantity necessary to melt ice, is equal to that which would raise water 150° of the new scale. The results may be seen in the following table.

1lb. hydrogen takes	7lbs.	oxygen,	prod.	8 lbs:	water,	melts 320lbs:	ic
—— carbur. hydrogen,	4	——	——	5	w: & car. acid	85	——
—— olefiant gas,	3.5	——	——	4.5	——	88	——
—— carbonic oxide,	.58	——	——	1.58	carb. acid	25	——
—— oil, wax and tal.	3.5	——	——	4.5	w. & car. ac.	104	——
—— oil of turp.	—	——	——	—	——	60	——
—— alcohol,	—	——	——	—	——	58	——
—— ether,	3	——	——	4	——	62	——
—— posphorus	1.5	——	——	2.5	phos. acid	60	——
—— charcoal	2.8	——	——	3.8	carb. acid	40	——
—— sulphur	—	——	——	—	sulph. acid	20	——
—— camphor	—	——	——	—	w. & car. ac.	70	——
—— caoutchouc	—	——	——	—	——	42	——

Lavoisier has left us a similar table derived from experiments on the calorimeter, for hydrogen, phosphorus, charcoal, oil and wax; and Crawford for hydrogen, charcoal, oil and wax, derived from their combustion in another apparatus. By reducing Crawford's results to a comparative scale with Lavoisier's, they will both appear as follows:

		according to Lavosier.	according to Crawford.
1lb Hydrogen by combustion melts		295lbs. ice	480lbs. ice.
— Phosphorus	———	100 —	— —
— Charcoal	———	96.5 —	69 —
— Wax	———	133. —	97 —
— Oil	———	148 —	89 —

HYDROGEN. The near coincidence of Lavosier's result and mine is an argument in favour of their accuracy. Crawford, I think, must have overrated the heat produced; his method of determining it, by the explosion of the gases by electricity, seems however susceptible of precision, and ought to be repeated. The truth perhaps lies between the two.

PHOSPHORUS. Lavoisier's result, which is much greater than mine, must, I think, be too high. I suspect that 66 is as much as can be fairly inferred.

CHARCOAL. The inferiority of my results to those of Crawford is what might be expected.

Mine must necessarily be rather too low. But Lavoisier is in this as well as all the other articles, hydrogen excepted, unwarrantably too high. I think Crawford will be found too high; his experiments on the heat produced by the respiration of animals, support this supposition.

WAX AND OIL. Crawford's results are a little lower than mine, which they ought not to be, and are doubtless below the truth. Lavoisier's certainly cannot be supported. This great philosopher was well aware of the uncertainty of his results, and expresses himself accordingly. He seems not to have had an adequate idea of the heat of hydrogen gas, which contributes so much to the quantity given out by its combustion; he compares, and expects to find an equation, between the heat given out by burning wax, &c. and the heat given out by the combustion of equal weights of hydrogen and charcoal in their separate state; but this cannot be expected, as both hydrogen and charcoal in a state of combination must contain less heat than when separate, agreeably to the general law of the evolution of heat on combination.—In fact, both Crawford and Lavoisier have been, in some degree, led away by the notion, that oxygenous gas was the sole or principal source of the light and heat produced by combustion. This is the more remarkable of the former, after he had proved that hydrogenous gas, one of the most frequent and abundant combustibles, possessed nearly five times as much heat as the same weight of oxygenous gas. Azote, another combustible, possesses as high and probably higher specific heat than oxygen. Oil, wax, tallow, alcohol, &c. would be far from being low in the table of specific heat, provided a table were formed comprehending bodies of every class. Charcoal and sulphur rank but low in the table. Upon the whole then, we cannot adopt the language of Crawford, "that inflammable "bodies contain little absolute heat," and "that the heat which is produced by com-"bustion is derived from the air, and not from "the inflammable body." This language may be nearly right as applied to the ordinary combustion of charcoal and pitcoal; but cannot be so when applied universally to combustible bodies.

After these remarks it is almost unnecessary to add that the heat, and probably the light also, evolved by combustion, must be conceived to be derived both from the oxygen and the combustible body; and that each contributes, for aught we know to the contrary, in proportion to its specific heat before the combustion. A similar observation may be made upon the heat produced by the union of sulphur with the metals, and every other chemical union in which heat is evolved.

Before we conclude this section it may be proper to add, for the sake of those who are more immediately interested in the economy of fuel, that the heat given out by the combustion of 1lb. of charcoal, and perhaps also of pitcoal, is sufficient (if there were no loss) to raise 45 or 50 lbs. of water from the freezeing to the boiling temperature; or it is sufficient to convert 7 or 8 lbs. of water into steam. If more than this weight of coal be used, there is a proportionate quantity of heat lost, which ought, if possible, to be avoided.

SECTION 6.

ON THE

NATURAL ZERO OF TEMPERATURE;

Or absolute Privation of Heat.

If we suppose a body at the ordinary temperature to contain a given quantity of heat, like as a vessel contains a given quantity of water,

It is plain that by abstracting successively small equal portions, the body would finally be exhausted of the fluid. It is an object of primary importance in the doctrine of heat to determine, how many degrees of the ordinary scale of temperature a body must be depressed before it would lose all its heat, or become absolutely cold. We have no means of effecting this by direct experiment; but we can acquire data for a calculus, from which the zero may be approximated with considerable accuracy.

The data requisite for the calculus are the exact specific heats of the several bodies operated upon, and the quantity of heat evolved, or absorbed by bodies, in cases of their chemical combinations or otherwise. These data are not to be acquired without great care and circumspection; and hence the great diversity of the results hitherto obtained in this difficult investigation. According to some, the zero is estimated to be 900° below the common temperature; whilst, according to others, it is nearly 8000° below the same. These are the extremes; but various determinations of an intermediate nature are to be found.

The most simple case in theory is that of

ice and water: supposing the capacities of these two bodies to be as 9 to 10, at the temperature of 32°, it is known that ice of 32° requires as much heat as would raise water 150°, to convert it into water of 32°, or to melt it. Consequently, according to the 8th formula, page 57, water of 32°, must contain 10 times as much heat, or 1500°. That is, the zero must be placed at 1500° below the temperature of freezing water. Unfortunately, however, the capacity of ice has not been determined with sufficient accuracy, partly because of its being a solid of a bad conducting power, but principally because the degrees of the common thermometer below freezing, are very erroneous from the equal division of the scale.

Besides the one already mentioned, the principal subjects that have been used in this investigation are, 1st, mixtures of sulphuric acid and water; 2d, mixtures of lime and water; 3d, mixture or combination of nitric acid and lime; and 4th, combustion of hydrogen, phosphorus and charcoal. Upon these it will be necessary to enlarge.

Mixture of Sulphuric Acid and Water.

According to the experiments of Lavoisier

and Laplace on the calorimeter, a mixture of sulphuric acid and water in the proportion of 4 to 3 by weight, determines the zero at 7292° below freezing water, reckoning by Fahrenheit. But a mixture of 4 acid with 5 water, determines the same at 2630°.

Gadolin made several experiments on mixtures of sulphuric acid and water, the results of which are as accurate as can be expected in a first essay of the kind. He has not determined the zero from his experiments, but taking it for granted to be 1400° below the freezing point on the supposition that the capacities of ice and water are as 9 to 10, he has enquired how far his experiments corroborate the same, by comparing the capacities of the mixtures by experiment with those calculated from the previous assumption. His results are thus curtailed in their utility; but as he has given us data sufficient to calculate the zero from each experiment, it will be proper to see how far they accord with Lavoisier's, or those of others.

Taking the specific heat of water at 1, Gadolin finds, by direct experiment, the specific heat of concentrated sulphuric acid to be .339 (See Crawford on heat, page 465); he then mixes the acid and water in various proportions, observes the increase of temperature, and then finds the capacities of the mixtures. Whence we have data to find the zero by formula 9, page 58. In giving his numbers, I have changed his scale, the centigrade, to Fahrenheit's.

Acid	Water	heat evolv.	capa. of mix.	comp. zero
4	+ 1	194°	.442	2936°
2	+ 1	203	.500	1710
1	+ 1	161	.605	1510
1	+ 2	108	.749	2637
1	+ 5	51	.876	3230
1	+ 10	28	.925	1740

The mean of these is 2300°, which is far beyond what Gadolin supposes to be the zero, as deduced from the relative capacities of ice and water, and to which he seeks to accommodate these experiments.

As the heat evolved upon the mixture of sulphuric acid and water is so considerable, and as all three articles are liquids, and consequently admit of having their capacities ascertained with greater precision, I have long been occasionally pursuing the investigation of the zero from experiments on these liquids. The strongest sulphuric acid of 1.855, I find has the specific heat .33, and

Acid	Water	sp. gr.	heat evol.	capa. of mix.	zero
5.77	+ 1	(1.78)	160°	.420	6400°
1.6	+ 1	(1.520)	260	.553	4150
1	+ 2	(1.250)	100	.764	6000

I reject all mixtures where the heat is less than 100°, because the difference between the observed capacity of the mixture, and the mean capacity is too small to be determined with precision. These results differ materially from Gadolin's. I believe they will be found to be nearer approximations to the truth. When the two liquids are mixed in nearly equal weights, the results give the zero less remote than otherwise; this appears to be the case both with Gadolin and me; I have not yet been able to discover the cause of it; perhaps the capacity of such mixture increases with the temperature more than in the other cases.

Lime and Water.

Quicklime, that is, lime recently burned, has a strong affinity for water; when mixed in due proportion an intense heat is produced; the lime falls, or becomes slaked, and then may be denominated hydrat of lime. If no more water is put to quicklime than is sufficient to slake it, or pulverize it, three parts of lime, by weight, form four parts of hydrat, a perfectly dry powder, from which the water cannot be expelled under a red heat. If more water is added, the mixture forms mortar, a pasty compound, from which the excess of water may be expelled by a boiling heat, and the hydrat remains a dry powder. When hydrat of lime and water are mixed, no heat is evolved; hence the two form a mere mixture, and not a chemical compound. The heat then which is evolved in slaking lime, arises from the chemical union of three parts of lime and one of water, or from the formation of the hydrat, and any excess of water diminishes the sensible heat produced. Before any use can be made of these facts for determining the zero, it becomes necessary to determine the specific heat of dry hydrat of lime. For this purpose a given weight of lime is to be slaked with an excess of water; the excess must then be expelled by heat till the hydrat is $\frac{1}{3}$ heavier than the lime. A given weight of this powder may then be mixed with the same, or any other weight of water of another temperature, and its specific heat determined accordingly. By a variety of experiments made in this way, and with sundry variations, I find the specific heat of hydrat of lime about .40, and not .25 as in the table, page 62. Lime itself I find to be nearly .30. Crawford undervalues lime, by mixing cold lime with hot alcohol; the lime does not produce a sufficient effect on the alcohol, because it contains water, which acts upon the lime. I have no doubt a different specific heat would have been found, if cold alcohol had been poured on hot lime. The heat evolved in the formation of hydrat of lime may be found as follows: If 1 oz of lime be put into 4 oz. of water, the temperature of the mixture will be raised 100°; in this case $1\frac{1}{3}$ oz. hydrat is formed, and the heat evolved raises it together with $3\frac{2}{3}$ oz. water 100°; but $3\frac{2}{3}$ water contains 7 times the heat that $1\frac{1}{3}$ hydrat of lime does; therefore the heat given out is sufficient to raise 8 times the hydrat 100°, or once the hydrat 800°. Whence the heat evolved by mixing 3 parts of lime and 1 of water, is sufficient to raise the new compound 800°. Applying then the theorem in page 58, we obtain the zero = 4260° below the common temperature.

Nitric Acid and Lime.

According to the experiments of Lavoisier and Laplace, the specific heat of nitric acid, sp. gr. 1.3, is .661, and that of lime .217, and a compound of $9\frac{1}{3}$ parts of said acid, and one of lime, is .619. But supposing there was no change of capacity upon combination, this compound should only have the capacity .618; whereas, in fact, the mixture produces an increase of temperature of about 180°, and therefore ought to be found with a diminished capacity, or one below .618. Were this fact to be established, it would exhibit an inexplicable phenomenon, unless to those who adopt the notion of *free* caloric and *combined* caloric existing in the same body, or to speak more properly, of caloric combined so as to retain all its characteristic properties, and caloric combined so as to lose the whole of them. One error in this statement has already been pointed out, in regard to the capacity of lime. If we adopt the specific heat of lime to be .30, and apply the theorem for the zero, we shall find it to be 15770° below the common temperature, as deduced from the above data so corrected.

I took a specimen of nitric acid of the specific gravity 1.2, and found, by repeated trials, its specific heat to be .76 by weight. Into 4600 grains of this acid of 35° temperature, in a thin flask, 657 grains of lime were gradually dropped, and the mixture moderately agitated; in one or two minutes after 3-4ths of the lime was in and dissolved, the thermometer rose nearly to 212°, and the mixture was beginning to boil; it was suffered to cool 20°, when the rest of the lime was added, and it again rose to the boiling point; about 15 grains of insoluble residuum were left. These were taken out, and their place supplied by 15 grains of fresh lime, which were dissolved, and left a clear liquid nearly saturated, of 1.334 sp. gravity. The specific heat of this was found to be .69. The increase of temperature being called 200°, and the specific heat of lime being .30, we find the zero to be 11000° below the freezing temperature. The experiment was varied by taking acids of different strengths, and various proportions of lime, but the results still gave the zero more remote than either of the previous methods. Perhaps the reason may be that lime is still under-rated.

Combustion of Hydrogen.

Lavoisier finds the combustion of 1lb. of hydrogen to melt 295lbs. of ice. The results of my experience give 320lbs, and Crawford's 480.—Till this fact can be more accurately ascertained, we may take 400lbs. as approximating to the truth. Or, which amounts to the same thing, the combustion of 1lb. of hydrogen takes 7lbs. of oxygen, and gives out heat which would raise 8lbs. of water 7500°.

By adopting Crawford's capacities of hydrogen and oxygen, and applying the theorem, page 58, we find the zero 1290° from the common temperature. But if we adopt the preceding theory of the specific heat of elastic fluids, and apply the 4th corol. page 72, we must conclude that in the formation of steam, one half of the whole heat of both its elements is given out; the conversion of 8lbs of steam into water, will give out heat sufficient to melt 56lbs. of ice; therefore one half of the whole heat in 1lb. of hydrogen, and 7lbs. of oxygen together, or which is the same thing, the whole heat in 1lb. of hydrogen, or 7lbs. of oxygen separately, will melt 344lbs. of ice; now if from 688 we take 400, there remain 288 for the lbs. of ice, which the heat in 8lbs. of water, at the ordinary temperature, is sufficient to melt, or the heat in 1lb. is capable of melting 36lbs. of ice: hence the zero will be 5400° below freezing water.

Combustion of Phosphorus.

One pound of phosphorus requires 1½lb. of oxygen, and melts 66lbs. of ice. The specific heat of phosphorus is not known; but from analogy one may suppose it to have as much heat as oil, wax, tallow, &c. which is nearly half as much as water. From the last article it seems, that the whole heat in each lb. of oxygen is sufficient to melt 50lbs. of ice; whence the whole heat in both articles, previous to combustion, is sufficient to melt 75 + 18 = 93lbs. of ice. From which deducting 66, there remains 27 for the pounds of ice, which the heat in 2.5lbs. of phosphoric acid ought to melt. This would give the specific heat of that acid .30, a supposition not at all improbable. The result of the combustion of phosphorus seems then to corroborate that from hydrogen.

Combustion of Charcoal.

Crawford's data are, specific heat of charcoal .26, oxygen 4.749, carbonic acid 1.0454, and the heat given out by burning 1lb. of charcoal = 69lbs. ice = 10350°. It is now established beyond doubt, that 1lb. of charcoal requires 2.6lbs. of oxygen to convert it into carbonic acid. From these data, by the theorem, page 58, we deduce the zero = 4400° — But Crawford himself has not noticed this deduction. If we adopt the theory of specific heat, and the table founded on it, combined with the supposition of the zero being 6000° below the common temperature, (see pag 74) we shall have from the general formula, this equation,

$$\frac{(1+2.6) \times .491 \times h}{1 \times .26 + 2.6 \times 1.333 - 3.6 \times .491} = 6000^\circ$$

where h represents the degrees of temperature which the combustion of 1lb. of charcoal would raise the product, or 3.6lbs. of carbonic acid. From this, h is found = 6650°. But this heat would raise 3.6lbs. of water = 6650 × .491 = 3265°. Or it would raise 1lb. of water, 11750°; or it would melt 78lbs. of ice. Lavoisier finds the effect = 96lbs. and Crawford finds it = 69. So that the supposed distance of the zero is not discountenanced by the combustion of charcoal, as far as the theory is concerned.

Combustion of Oil, Wax and Tallow.

We do not know the exact constitution of these compounds, nor the quantity of oxygen which they require; but from the experiments of Lavoisier, as well as from some attempts of my own, I am inclined to think, that they are formed of about 5 parts of charcoal and 1 of hydrogen by weight, and that 6 parts require 21 of oxygen for their combustion, forming 19 parts of carbonic acid and 8 of water. Let it be supposed that the zero is 6900° below freezing water, or that the heat in water of 32°, is sufficient to melt 46lbs. of ice, then the heat in steam will be sufficient to melt 53lbs. By applying Cor. 1, at page 72, we shall find the heat in oxygenous gas = 60.5lbs. and in carbonic acid, 22.3lbs. The heat in 1lb. of oil, &c. equal to half that of water = 23lbs. which being added to 211.7, the heat in 3.5lbs. of oxygen, gives 234.7lbs. of ice, which would be melted by all the heat in 1lb. of oil and 3.5 of oxygen; but the products of combustion are 1.3lb. of water, and 3.2lbs. of carbonic acid, together containing as much heat as would melt 131.2lbs. of ice; this being subtracted from 234.7, leaves 103.5 for the ice to be melted by the heat evolved during the combustion of 1lb. of oil, wax or tallow, which agrees with the experiment. The conclusion then supports the supposition, that the zero is 6900° below freezing water.

Combustion of Ether, &c.

I have pretty accurately ascertained the products of the combustion of 1lb. of ether to be 1.75 water, and 2.25 carbonic acid, derived from its union with 3lbs. of oxygen. By instituting a calculation similar to the above, but on the supposition of the zero being 6000° below freezing water, I find the heat given out on the combustion of ether, ought to be = 67lbs. of ice: it was observed to be 62, and the difference may well be attributed to the loss unavoidable in my method of observation.

I might here enquire into the results of the combustion of the other articles mentioned in the table, page 78, as far as they affect the present question; but I consider those above noticed as the most to be depended upon. From the result of olefiant gas we may learn, that a combustible body in the gaseous state, does not give out much more heat than when in a liquid state; for, oil and olefiant gas certainly do not differ much in their constitution; one would therefore have expected the same weight of olefiant gas to have yielded more heat than oil, because of the heat required to maintain the elastic state; but it should seem that the heat requisite to convert a liquid to an elastic fluid, is but a small portion of the whole, a conclusion evidently countenanced by the experiments and observations contained in the preceding pages.

It may be proper now to draw up the results of my experience, reported in the present section, into one point of view.

			Zero below 32° Fahrenheit.
From a mixture of 5.77 sulphuric acid	and 1 water		6400°
——— 1.6 ———	1 ——		4150
——— 1 ———	2 ——		6000
——— 3 lime	1 ——		4260
——— 7 nitric acid	1 lime		11000
From the combustion of hydrogen			5400
——— phosphorus			5400
——— charcoal			6000
——— oil, wax and tallow			6900
——— ether			6000

The mean of all these is 6150. We are authorised then, till something more decisive appear, to consider the natural zero of temperature as being about 6000° below the temperature of freezing water, according to the divisions of Fahrenheit's scale. The differences of the above results are not greater than what may be ascribed to inaccuracies, except the 2d and 5th. I believe it will be impossible to reconcile these two to each other, unless it is upon the supposition of a change of capacity with change of temperature in one or both of the mixtures. This deserves farther enquiry.

Heat produced by Percussion and Friction

The heat produced by the percussion and friction of solid bodies, arises from one and the same cause, namely, from a condensation of volume, and consequent diminution of capacity of the excited body; exactly in the same manner as the condensation of air produces heat. It is a well known fact, that iron and other metals, by being hammered, become hot and condensed in volume at the same time; and if a diminution of capacity has not been observed it is because it is small, and has not been investigated with sufficient accuracy. That a change of capacity actually takes place cannot be doubted, when it is considered, that a piece of iron once hammered in this way, is unfitted for a repetition of the effect, till it has been heated in a fire and cooled gradually. Count Rumford has furnished us with some important facts on the production of heat by friction. He found that in boring a cannon for 30 minutes, the temperature was raised 70°; and that it suffered a loss of 837 grains by the dust, and scales torn off, which amounted to $\frac{1}{948}$ part of the cylinder. *On the supposition that all the heat was given out by these scales*, he finds they must have lost 66360° of temperature; when at the same time he found their specific heat not sensibly diminished. But this is manifestly an incorrect view of the subject: the heat excited does not arise from the scales merely, else how should hammering make a body red hot without any loss of scales? The fact is, the whole mass of metal is more or less condensed by the violence used in boring, and a rise of temperature of 70 or 100° is too small to produce a sensible diminution in its capacity for heat. Does Count Rumford suppose, that if in this case the quantity of metal operated upon had been 1lb. and the dust produced the same as above, that the whole quantity of heat evolved would have been the same?

The phenomena of heat produced by friction and percussion, however, sufficiently shew that the zero of temperature cannot be placed at so small a distance as 1000° or 1500° below the common temperature, as has been determined by some philosophers.

SECTION 7.

ON THE

MOTION AND COMMUNICATION OF HEAT,

Arising from inequality of Temperature.

As from various sources the temperature of bodies is liable to perpetual fluctuation, it becomes of importance to determine the nature of the motion of heat in the same body, and in its passage from one body to another, arising from its incessant tendency to an equilibrium.

A solid bar being heated at one end, and exposed to the air, the heat is partly dissipated in the air, and partly conducted along the bar, exhibiting a gradation of temperature from the hot to the cold end. This power of conducting heat varies greatly, according to the nature of the subject: in general, metals, and those bodies which are good conductors of electricity, are likewise good conductors of heat; and *vice versâ*.

When a fluid is heated at its surface, the heat gradually and slowly descends in the same manner as along a solid; and fluids seem to have a difference in their conducting power analogous to that of solids. But when the heat is applied to the bottom of a vessel, containing a fluid, the case is very different; the heated particles of the fluid, in consequence of their diminished specific gravity, form an ascending current and rise to the surface, communicating a portion of heat in their ascent to the contiguous particles, but still retaining a superiority of temperature; so that the increase of temperature in the mass is first observed at the surface, and is constantly greatest there till the commencement of ebullition in liquids, at which period the temperature is uniform. The conducting power of fluids then arises from two distinct sources; the one is the same as in solids, namely, a gradual progress of the heat from particle to particle, exclusive of any motion of the particles themselves; the other arises from the internal motion of the particles of the fluid, by which the extremes of hot and cold are perpetually brought into contact, and the heat is thus diffused with great celerity. The latter source is so much more effectual than the former, that some have been led, though without sufficient reason, to doubt the existence of the former, or that fluids do convey heat in the same manner as solids.

Nothing appears, then, but that the communication of heat from particle to particle, is performed in the same way in fluids as in solids; the rapidity of its diffusion in fluids, is to be ascribed to an hydrostatical law. But there is another method by which heat is propagated through a vacuum, and through elastic fluids, which demands our particular notice. By this we receive the heat of the sun; and by this, when in a room, we receive the heat of an ordinary fire. It is called the *radiation* of heat; and the heat, so propelled, is called *radiant heat.*

Till lately we have been used to consider the light and heat of the sun as the same thing. But Dr. Herschel has shewn, that there are rays of heat proceeding from the sun, which are separable by a prism from the rays of light; they are subject to reflection, like light; and to refraction, but in a less degree, which is the cause of their separability from light. The velocity of radiant heat is not known; but it may be presumed to be the same as that of light, till something appears to the contrary. An ordinary fire, red hot charcoal, or indeed any heated body, radiates heat, which is capable of being reflected to a focus, like the light and heat of the sun; but it should seem to be not of sufficient energy to penetrate glass, or other transparent bodies so as to be refracted to an efficient focus.

Several new and important facts relative to the radiation of heat, have lately been ascertained by Professor Leslie, and published in his "Enquiry on Heat." Having invented an ingenious and delicate air thermometer, well adapted for the purpose, he was enabled to mark the effects of radiation in a great variety of cases and circumstances, with more precision than had previously been done. Some of the principal facts respecting the radiation of heat, which have either been discovered or confirmed by him, it will be proper to mention.

1. If a given vessel be filled with hot water, the quantity of heat which radiates from it, depends chiefly upon the nature of the exterior surface of the vessel. Thus, if a canister of tinned iron be the vessel, then a certain quantity of heat radiates from it; if the said vessel be covered with black paint, paper, glass, &c. it will then radiate 8 times as much heat in like circumstances.

2. If the bulb of the thermometer be covered with tinfoil, the impression of the radiant heat is only $\frac{1}{5}$th of that upon the glass surface.

3. A metallic mirror reflects 10 times as much heat from an ordinary fire, or from any heated body, as a similar glass mirror does. This last is found to reflect the heat from its *anterior* surface, and not from the quicksilvered one, which is the most essential in reflecting solar light and heat. Here then is a striking difference between solar and culinary heat.

From these facts it appears, that metals and other bodies which are eminently disposed to *reflect* radiant heat, are not disposed to *absorb* it in any remarkable degree; whereas, black paint, paper, glass, &c. are disposed to *absorb* it, and consequently to *radiate* it again in proper circumstances.

4. Screens of glass, paper, tinfoil, &c. being placed between the radiating body and the reflector, were proved to intercept the radiant heat completely; but being heated themselves by the direct radiant heat, in time the thermometer was affected by their radiation.—The heat radiating from hot water, does not then seem capable of being transmitted through glass, like the solar heat.

5. Radiant heat suffers no sensible loss in its passage through the air; a greater or less radiant body produces the same effect, provided it subtends the same angle at the reflector, agreeing with light in this respect.

6. The intensity of reflected heat diminishes inversely as the distance; whereas, in light, it is the same at all distances; the focus of heat too differs from that of light; it is nearer the reflector; the heating effect diminishes rapidly in going outwards, but slowly in going inwards towards the reflector.—This seems to intimate the want of perfect elasticity in radiant heat.

7. A hollow globe of tin, four inches in diameter, being filled with hot water, cooled

from 35° to 25° centigrade in 156 minutes; the same painted with lamp-black, cooled from 35° to 25° in 81 minutes. The air of the room was 15°.

8. When a heated body is whirled through the air, the additional cooling effect is directly proportional to the velocity.

9. In air the rate of cooling of a hollow glass globe filled with hot water, and that of the same globe covered with tinfoil, is not constant at all temperatures. The disproportion is greater in low temperatures, and less in high. Thus, in the present case, Mr. Leslie finds the variable ratio to be as $105 + h$ for glass, and as $50 + h$ for tin, where h represents the elevation of temperature in degrees. According to this the rate of cooling of a vitreous and a metallic surface is nearly the same at very high temperatures; but is nearly as 105 to 50, when h is very little.—No differences are observed in their rates of cooling in water.

10. After a long and intricate, but ingenious investigation, Mr. Leslie finds the cooling power of the air upon a hollow sphere, six inches in diameter, and filled with boiling water, to be as follows: namely, in each minute of time the fluid loses the following

fractional parts of its excess of temperature, by the three distinct sources of refrigeration in the air undermentioned:

By abduction, that is, the proper conducting power of air, the 524th.

By recession, that is, the perpendicular current of air excited by the heated body, the $h \times 21715$th.

By pulsation, or radiation, the 2533d part from a metalic surface, and eight times as much, or the 317th part from a surface of paper; (It should be observed, that Mr. Leslie contends that air is instrumental in the radiation of heat, which is contrary to the received opinion.)

11. A body cools more slowly in rarefied air, than in air of the common density: and the different species of air have their respective refrigerating powers. Common air and hydrogenous gas exhibit remarkable differences. According to Mr. Leslie, if the cooling power of common air upon a vitreous surface be denoted by unity, that of hydrogenous gas will be denoted by 2,2857; and upon a metallic surface the ratio is .5 to 1.7857. In common air the loss from a vitreous surface is .57 by radiation, and .43 by the other two causes: from a metallic surface, .07 and .43. In hydrogenous gas the loss from a vitreous surface is .57

by radiation, and 1.71 by the other causes; from a metallic surface, .07 and 1.71.—He finds the radiation to be the same in the two gases, and to be very little diminished by rarefaction; but the effects of the other refrigerating powers rapidly diminish with the density.

Those who wish to see the experiments and reasonings from which these important conclusions are derived, must have recourse to Mr. Leslie's work: but as some of the facts and opinions appear from my experience to be questionable, I shall now proceed to state what has occurred to me on these subjects.—I have no reason to withhold my assent from the first 8 articles; but the last 3 are not equally satisfactory.

Before we enter upon a detail of experiments, it will be proper to point out the correspondence of the new thermometric scale with the old one in the higher parts, it being only given briefly in the table, page 14.

Correspondences of the Thermometric Scales.

old scale.	new scale.	old scale.	new scale.
212°	212°	409°.8	342°
225	222	427. 3	352
238.6	232	445. 3	362
252.6	242	463. 6	372
266.8	252	482. 2	382
281.2	262	501.	392
296.2	272	520. 3	402
311.5	282	539. 7	412
327.	292	559. 8	422
342.7	302	580. 1	432
359.2	312	600. 7	442
375.8	322	621. 6	452
392.7	332	642	762

Experiment 1.

A mercurial thermometer having a bulb of half an in inch in diameter, and a scale of about 8 inches long from freezing to boiling mercury, was heated to 442° new scale, and suffered to cool in a horizontal position in air of 42°. The bulb in this and every other instrument projected several inches below the scale. The times of cooling were the same from 442° to 242°, from 242° to 142°, and from 142° to 92°, namely, 2 minutes and 20 seconds each. This was often repeated; the times of cooling were always within 4 or 5 seconds of that above, and when any differences in the successive intervals took place, the times were always observed to be rather less in the higher, and more in the lower parts of the scale.

From this experiment it appears, that the thermometer was raised 400° above the temperature of the air, or to 600° of the old scale; it lost 200° of temperature in the first interval of time, 100° in the second, and 50° in the third. This result goes to establish the principle announced at page 12, that, according to the new graduation, *the temperature descends in geometrical progression to equal increments of time.*

Experiment 2.

According to Mr. Leslie, the same law of cooling does not take place from a metallic as from a vitreous surface; this always appeared to me very surprising, and I was anxious to satisfy myself more particularly as to the fact. With this view, I took another mercurial thermometer, with a bulb of .7 inch diameter, and scale of 12 inches, having a range from 0 to 300° old scale, and corresponding new scale attached to it. This was heated, and the times of cooling through every successive 10 degrees of the new scale were noticed repeatedly; the bulb was then covered with tinfoil, pasted upon it, and the surface made as smooth as well could be ; the thermometer was then heated, and the times of cooling were again noticed as before, repeatedly. The mean results follow ; and a column of the differences of the logarithms of the degrees expressing the elevation of temperature above that of the surrounding air, which was 40°. The temperature of the thermometer was raised to 275° per scale ; that is, 235° above the air, and it is obviously most convenient to reckon from the temperature of the air considered as zero : in which case 19 represents the difference of the logarithms of 235 and 225, &c.

Thermom. cooled.				Bulb clear. Seconds.	Bulb cov. with tinfoil. Seconds.	Dif. of Logarith.
From 235°	to	225°	in	11	17	19
225	to	215		12	18	20
215	to	205		13	18	21
205	to	195		14	19	22
195	to	185		15	20	23
185	to	175		16	22	24
175	to	165		17	24	25
165	to	155		19	26	27
155	to	145		20	28	29
145	to	135		21	30	31
135	to	125		22	31	33
125	to	115		24	33	36
115	to	105		27	36	39
105	to	95		30	40	43
95	to	85		34	48	48
85	to	75		39	54	54
75	to	65		45	62	62
65	to	55		52	73	73
55	to	45		62	88	87
45	to	35		78	110	109
35	to	25		120	165	146
25	to	15		160	244	222
				851	1206	1193

By inspecting this table, it appears that the whole time of cooling when the bulb was clear was 8.51 seconds, and when covered with tinfoil was 1206 seconds, which numbers are nearly as 17 to 24. But the times in cooling from 175° to 165° were 17 and 24 seconds respectively; and the times in cooling from 95° to 85° were 34 and 48 respectively, which are exactly in the ratio of the whole times: and by examining any two corresponding times, they will be found to be as 17 to 24 nearly. Whence it follows that the same law of progressive cooling applies to a metallic as to a vitreous surface, contrary to the results of Mr. Leslie's experience. It must not however be understood that this ratio for the two kinds of surfaces is quite correct; however carefully the bulb of a thermometer may be coated with tinfoil, the surface is necessarily enlarged, which makes it cool more quickly than if the metallic surface were the very same quantity as the vitreous.

The differences of the logarithms happening accidentally so nearly to coincide in magnitude with the times of cooling of the metallic surface, they require no reduction, and we have an opportunity of seeing how far the law of geometrical progression in cooling is supported by this experiment. It appears that for 5 or 6 of the highest intervals of temperature, the times of cooling were rather smaller, and for the two last rather larger than required by the law.

Experiment 3.

As Mr. Leslie found the times of cooling of metallic surfaces considerably enlarged, in moderate elevations of temperature more especially, I took another thermometer having a smaller bulb, and a scale of an inch for 10 degrees, this was treated as in the last experiment, and the results were as under:

Thermom. cooled.					Bulb clear. second	Bulb coated with tinfoil. seconds	Log. ratios. reduced.
From	75°	to	65°	in	38	46	46
	65	to	55		46	55	54
	55	to	45		54	64	65
	45	to	35		65	78	81
	35	to	25		86	103	109
	25	to	15		130	158	165
	15	to	5		310	370	355
					729	874	875

Here the whole times of cooling, and the several parts are almost accurately as 10 for the vitreous, and as 12 for the metallic surface. They very nearly accord too with the logarithmic ratios. The effect of the metallic surface differs less from that of the vitreous in this than in the former experiment; because the bulb being smaller, it was more than proportionally increased in surface by the tinfoil, which was pasted on in small slips, and consequently was twofold in many places.

Being from these results pretty well satisfied that the surfaces of bodies do not disturb the law of their refrigeration, though they materially affect the time, yet in consequence of the general accuracy of Mr. Leslie's experiments, I was desirous to ascertain the results in his own way, more particularly because for the reason assigned above, my method did not give the true rates of cooling of *equal* surfaces.

Experiment 4.

I took two new tin canisters, such as are commonly used for tea, of a cylindrico-conical shape, and each capable of holding 15 oz. of water The surface of one of them was covered with brown paper pasted on it; instead of the usual lid, a cork of $1\frac{1}{2}$ inch. diameter was adapted to both, and through a hole in the centre of this, the tube of a delicate thermometer was inserted, with a scale of the new graduation affixed above the cork. Both canisters were contrived to be suspended by small strings when filled with water, and to have the thermometer with its bulb in their centers. They were successively filled with boiling water, and suspended in the middle of a room of the temperature 40°, and the times of cooling through each successive 10 degrees were noticed as below.

Water cooled.				Canister covered with paper.	Naked canister.	Logarith. ratios.
From 205°	to	195°	in	6.5 min.	10 min.	11
195	to	185		7	10.5	12
185	to	175		7.5	11+	13
175	to	165		8+	12	13
165	to	155		9	13.5	14
155	to	145		10	15	16
145	to	135		11.5	17	17
135	to	125		13	19	19
125	to	115		14.5	21.5	22
115	to	105		16	24	25
105	to	95		20	30	29
95	to	85		25	38	35
85	to	75		31	46	44
75	to	65		40	60	60
				219	327.5	330

Here the results are equally satisfactory and important; not only the times of cooling are in the uniform ratio of 2 to 3 throughout the range; but they almost exactly accord with the logarithmic ratios, indicating the geometric progression in cooling. As experiments of this sort are capable of being repeated by any one without the aid of any expensive instrument or any extraordinary dexterity; it will be unnecessary to insist upon the accuracy of the above. It will be understood that the range of cooling was from 205° of the new scale, to 65° of the same, the air being 40°, or 25° below the extremity of the range, which corresponds with 57° of the old scale.

It will be proper now to enquire into the cause of the difference in the times of cooling arising from the variation of surface. Mr. Leslie has shewn the surface has no influence upon the time of cooling when immersed in water; it should seem then that the difference of surfaces in the expenditure of heat arises from their different powers of radiation solely; indeed Leslie has proved by direct experiments that the heat radiating from a vitreous or paper surface is 8 times as great as that from a metallic surface. Taking this for granted, we can easily find the portions of heat dispersed by radiation, and conducted away by the atmosphere. For, let 1 denote the quantity of heat conducted away by the atmosphere, from a vitreous or metallic surface in any given small portion of time, and x the quantity radiated from a metallic surface in the same time; then $8x$ will be the quantity radiated from a vitreous surface in that time; and from the result of the last experiment we shall have, $2 : 3 :: 1+x : 1 + 8x$; whence $2 + 16x = 3 + 3x$, and

$x = \frac{1}{13}$; this gives $1\frac{1}{13}$, for the whole heat discharged by metal, and $1\frac{8}{13}$ for that discharged by glass in the same time, where the unit expresses the part conducted, and the fraction the part radiated.

That is, from a metallic surface 13 parts of heat are conducted away by the air and 1 part radiated; from a vitreous surface 13 parts are conducted, and 8 parts radiated, in a given time.

The quantity of heat discharged by radiation from the most favourable surface, therefore, is probably not more than .4 of the whole, and that conducted away by the air not less than .6.—Mr. Leslie however deduces .57 for the former, and .43 for the latter; because he found the disproportion in the times of cooling of vitreous and metallic surfaces greater than I find it in the lower part of the scale.

The obvious consequences of this doctrine in a practical sense are,

1. In every case where heat is required to be retained as long as possible, the containing vessel should be of metal, with a bright clear surface.

2. Whenever heat is required to be given out by a body with as much celerity as possible, the containing vessel, if of metal, ought to be painted, covered with paper, char-

coal, or some animal or vegetable matter; in which case the heat given out will be 3 parts for 2 from a metallic surface.

Refrigeration of Bodies in various Kinds of Elastic Fluids.

Bodies cool in very different times in some of the elastic fluids. Mr. Leslie was the first, I believe, who noticed this fact; and he has given us the results of his experiments on common air and hydrogenous gas, of the common density, and also rarefied in various degrees.— I made some experiments with a view to determine the relative cooling powers of the gases, the results of which it may be proper to give. My apparatus was a strong phial, containing about 15 or 20 cubic inches; a perforated cork containing the stem of a thermometer was adapted to it, so as to be air tight; two marks were made with a file on the tube of the thermometer, comprizing an interval of 15 or 20°, about blood heat. The bottle was filled with any proposed gas, and after it had acquired the temperature of the surrounding air, the stopper was withdrawn, and the heated thermometer with its cork was instantly inserted; the number of seconds which elapsed whilst the mercury descended from the upper to the under mark were then noted, as under. The surrounding air was of a constant temperature.

Thermometer immersed	cooled in
In carbonic acid gas	112 seconds.
— sulphuretted hydrogen, nitrous oxide, and olefiant gas	100 +
— com. air, azotic and oxyg. gas	100
— nitrous gas	90
— carburet. hyd. or coal gas	70
— hydrogen	40

The refrigerating effect of hydrogen is truly remarkable; I cooled the thermometer 10 times successively in a bottle of hydrogen gas; at each experiment the instrument was taken out, and the stopper put in, till the original temperature was restored; by this, a portion of the hydrogen escaped each time, and an equal portion of common air was admitted; the times of cooling regularly increased as follows; viz. 40, 43, 46, 48, 51, 53, 56, 58, 60 and 62 seconds, respectively; at this time the mixture was examined, and found half hydrogen and half common air. Equal measures of hydrogen and common air were then mixed together, and put into the bottle, and the heated thermometer was found to cool from mark to mark, in 62 seconds as before.

Condensed air cools bodies more rapidly than air of common density; and rarefied air less rapidly, whatever be the kind.—The results of my own experience for common air were as follows:

Density of the air.	Therm. cools in
2	85 seconds.
1	100
$\frac{1}{2}$	116
$\frac{1}{4}$	128
$\frac{1}{8}$	140
$\frac{1}{16}$	160
$\frac{1}{32}$	170

A small receiver of hydrogen gas, which cooled the thermometer in 40 seconds, when rarefied 7 or 8 times, took 70 seconds to cool the same. But the exact effects of rarefaction on this and the other gases were not determined.

From Mr. Leslie, we learn that in hydrogenous gas, there is little difference between the time of cooling of a vitreous and metallic surface, the former being as 2.28, and the latter as 1.78, from which he justly infers "this inequality of effect [between atmospheric air and hydrogenous gas] proves its influence to be exerted chiefly, if not entirely, in augmenting the abductive portion."

The expenditure of heat by radiation being the same in hydrogenous gas as in atmospheric air, we may infer it is the same in every other species of gas; and therefore is performed independently of the gas, and is carried on the same in vacuo as in air. Indeed Mr. Leslie himself admits that the diminution of the effect consequent upon rarefaction is extremely small, which can scarcely be conceived if air were the medium of radiation.

The effect of radiation being allowed constant, that of the density of the air may be investigated, and will be found, I believe, to vary nearly or accurately as the cube root of the density. In order to compare this hypothesis with observation, let 100 = time of cooling in atmospheric air, the density being 1; then from what has been said above, .4 will represent the heat lost by a vitreous surface by radiation, and .6 that lost by the conducting power of the medium. Let t = the time of cooling in air of the density d; then if $100 : .4 :: t : .004\,t$ = the heat lost by radiation; but the heat conducted away is, by hypothesis, as the time

× by the cube root of the density = .006 $t\sqrt[3]{d}$; whence $.004t + .006\,t\sqrt[3]{d} = 1$, and

$$t = \frac{1}{.004 + .006\sqrt[3]{d}}$$

Calculating from this formula, we shall find the times of cooling in common air of the several densities as under:

Density of the air.		Times of cooling.
2	....	86.5 seconds.
1	....	100
$\frac{1}{2}$	....	114
$\frac{1}{4}$	....	129
$\frac{1}{8}$	....	143
$\frac{1}{16}$	....	157
$\frac{1}{32}$	....	170
$\frac{1}{64}$	....	182
$\frac{1}{128}$	....	193
$\frac{1}{\text{infinity.}}$	....	250

This table accords nearly with the preceding one, the result of actual observation.—In the same way might the times of cooling of a metallic surface in rarefied air be found, by substituting .0007 for .004, and .0093 for .006 in the preceding formula.

The cooling power of hydrogenous gas independent of radiation, may be found thus:

if 100″ : .4 :: 40″ : .16 = the heat lost by radiation in that gas in 40 seconds; whence .84 = the heat conducted away by the air in 40″, or .021 per second; but in common air the loss per second by abduction is only .006; from this it appears that the refrigerating power of hydrogenous gas is $3\frac{1}{2}$ times as great as that of common air.

It may be asked what is the cause why different gases have such different cooling effects, especially on the supposition of each atom of all the different species possessing the same quantity of heat? To this we may answer that the gases differ from each other in two essential points, in the number of atoms in a given volume, and in the weight or inertia of their respective atoms. Now both number and weight tend to retard the motion of a current: that is, if two gases possess the same number of particles in a given volume, it is evident that one will disperse heat most quickly which has its atoms of the least weight; and if other two gases have particles of the same weight, that one will most disperse heat which has the least number in a given volume; because the resistance will be as the number of particles to be moved, in like circumstances. Of the gases that have nearly the same number of particles in the same volume, are, hydrogen, carburetted hydrogen, sulphuretted hydrogen, nitrous oxide, and carbonic acid. These conduct heat in the order they are written, hydrogen best and carbonic acid worst; and the weights of their ultimate particles increase in the same order (see page 73). Of those that have their atoms of the same weight and their number in a given volume diff rent, are oxygen and carburetted hydrogen: the latter has the greater cooling power and the fewer particles in a given volume.

SECTION 8.

ON THE TEMPERATURE OF THE ATMOSPHERE.

It is a remarkable fact, and has never, I believe, been satisfactorily accounted for, that the atmosphere in all places and seasons is found to decrease in temperature in proportion as we ascend, and nearly in an arithmetical progression. Sometimes the fact may have been otherwise, namely, that the air was colder at the surface of the earth than above, particularly at the breaking of a frost, I have observed it so; but this is evidently the effect of great and extraordinary commotion in the atmosphere, and is at most of a very short duration. What then is the occasion of this diminution of temperature in ascending? Before this question can be solved, it may be proper to consider the defects of the common solution.—Air, it is said, is not heated by the direct rays of the sun; which pass through it as a transparent medium, without producing any calorific effect, till they arrive at the surface of the earth. The earth being heated, communicates a portion to the contiguous atmosphere, whilst the superior strata in proportion as they are more remote, receive less heat, forming a gradation of temperature, similar to what takes place along a bar of iron when one of its ends is heated.

The first part of the above solution is probably correct: Air, it should seem, is singular in regard to heat; it neither receives nor discharges it in a radiant state; if so, the propagation of heat through air must be effected by its conducting power, the same as in water. Now we know that heat applied to the under surface of a column of water is propagated upwards with great celerity, by the actual ascent of the heated particles: it is equally certain too that heated air ascends. From these observations it should follow that the causes assigned above for the gradual change of temperature in a perpendicular column of the atmosphere, would apply directly to a state of temperature the very reverse of the fact; namely, to one in which the higher the ascent or the more remote from the earth the higher should be the temperature.

Whether this reasoning be correct or not, it must I think be universally allowed, that the fact has not hitherto received a satisfactory explanation. I conceive it to be one involving a new principle of heat; by which I mean a principle that no other phenomenon of nature presents us with, and which is not at present recognized as such. I shall endeavour in what follows to make out this position.

The principle is this: *The natural equilibrium of heat in an atmosphere, is when each atom of air in the same perpendicular column is possessed of the same quantity of heat;* and consequently, *the natural equilibrium of heat in an atmosphere is when the temperature gradually diminishes in ascending.*

That this is a just consequence cannot be denied, when we consider that air increases in its capacity for heat by rarefaction: when the quantity of heat is given or limited, therefore the temperature must be regulated by the density.

It is an established principle that any body on the surface of the earth unequally heated is observed constantly to tend towards an equality of temperature; the new principle announced above, seems to suggest an exception to this law. But if it be examined, it can scarcely appear in that light. *Equality of heat and equality of temperature,* when applied to the same body in the same state, are found so uniformly to be associated together, that we scarcely think of making any distinction between the two expressions. No one would object to the commonly observed law being expressed in these terms: *When any body is unequally heated, the equilibrium is found to be restored when each particle of the body becomes in possession of the same quantity of heat.* Now the law thus expressed is what I apprehend to be the true general law, which applies to the atmosphere as well as to other bodies. It is an *equality of heat,* and not *an equality of temperature* that nature tends to restore.

The atmosphere indeed presents a striking peculiarity to us in regard to heat: we see in a perpendicular column of air, a body without any change of form, slowly and gradually changing its capacity for heat from a less to a greater; but all other bodies retain a uniform capacity throughout their substance.

If it be asked why an equilibrium of heat should turn upon the equality in *quantity* rather than in *temperature*; I answer that I do not know: but I rest the proof of it upon the fact of the inequality of temperature observed in ascending into the atmosphere. If the natural tendency of the atmosphere was to an equality of temperature, there does not appear to me any reason why the superior regions of the air should not be at least as warm as the inferior.

The arguments already advanced on behalf of the principle we are endeavouring to establish, are powerfully corroborated by the following facts:—By the observations of Bouguer, Saussure, and Gay Lussac, we find that the temperature of the air at an ele ation where its weight is $\frac{1}{2}$ that at the surface, is about 50° Fahrenheit less than that at the surface: and from my experiments (Manch. Mem. vol. 5. page 525.) it appears that air being suddenly rarefied from 2 to 1 produces 50° of cold. Whence we may infer, that a measure of air at the earth's surface being taken up to the height above-mentioned, preserving its original temperature, and suffered to expand, would become two measures, and be reduced to the same temperature as the surrounding air; or *vice, versâ,* if two measures of air at the proposed height were condensed into one measure, their temperature would be raised 50°, and they would become the same in density and temperature, as the like volume of air at the earth's surface. In like manner we may infer, that if a volume of air from the earth's surface, to the summit of the atmosphere were condensed and brought into a horizontal position on the earth's surface, it would become of the same density and temperature as the air around it, without receiving or parting with any heat whatever.

Another important argument in favour of the theory here proposed may be derived from the contemplation of an atmosphere of vapour. Suppose the present ærial atmosphere were to be annihilated, and one of steam or aqueous vapour were substituted in its place; and suppose further, that the temperature of this atmosphere at the earth's surface were every where 212° and its weight equal to 30 inches of mercury. Now at the elevation of about 6 miles the weight would be 15 inches or $\frac{1}{2}$ of that below, at 12 miles, it would be 7.5 inches, or $\frac{1}{4}$ of that at the surface, &c. and the temperature would probably diminish 25° at each of those intervals. It could not diminish more; for we have seen (page 14) that a diminution of temperature of 25° reduces the force of vapour one half; if therefore a greater reduction of temperature were to take place, the weight of the incumbent atmosphere would condense a portion of the vapour into water, and the general equilibrium would thus be disturbed perpetually from condensations in the upper regions. But if we suppose on the other hand, that the diminution of temperature in each of these intervals is less than 25°, then the upper regions could admit of more vapour without condensation; but it must take place at the surface, because vapour at 212° cannot sustain more than the weight of 30 inches of mercury.

These three supposed cases of an aqueous vapour atmosphere may be otherwise stated thus:

1. The specific gravity of steam at the earth's surface being supposed .6 of atmospheric air, and the weight of the atmosphere of steam equal to 30 inches of mercury, its temperature at the surface would be 212°; at 6 miles high, 187°; at 12 miles, 162°; at 18 miles, 137°; at 24 miles, 112°, &c.—In this case the density, not only at the surface, but every where, would be a maximum, or the greatest possible for the existing temperature; so that

a perfect equilibrium having once obtained, there could be neither condensation nor evaporation in any region. For every 400 yards of elevation, the thermometer would descend 1 degree.

2. If the atmosphere were constituted just as above, except that the temperature now diminished more rapidly than at the rate of 25° for 6 miles; then the temperature of the higher regions not being sufficient to support the weight, a condensation must take place; the weight would thus be diminished, but as the temperature at the surface is always supposed to be kept at 212°, evaporation must go on there with the design to keep up the pressure at 30 inches. Thus there would be perpetual strife between the recently raised vapour ascending, and the condensed drops of rain descending. A position much less likely than the preceding one.

3. The same things being supposed as before, but now the temperature decreases more slowly than at the rate of 25° for 6 miles: in this case the density of the steam at the earth's surface would be a maximum for the temperature, but no where else; so that if a quantity of water were taken up to any elevation it would evaporate; but the increased weight of the atmosphere would produce a condensation of

steam into water on the ground. In this case then there would not be that equilibrium, which we see in the 1st case, and which accords so much more with the regularity and simplicity generally observable in the laws of nature.*

* I owe to Mr. Ewart the first hint of the idea respecting elastic fluids, which I have endeavoured to expand in the present section; he suggested to me some time ago, that it was probable steam of any low temperature, as 32°, of maximum density, contained the same quantity of absolute heat as the like weight of steam of 212° of maximum density; and that consequently if it could be gradually compressed without losing any heat, that is, if the containg vessel kept pace with it in increase of temperature, there would never be any condensation of steam into water, but it would constantly retain its elasticity.

In fact the heat (1000°), which is given out by steam when it is condensed into water, is merely heat of compression; there is no change in the affinity of the molecules of water for heat; the expulsion is occasioned solely by the approximation of the molecules, and would be precisely the same whether that approximation was occasioned by external compression or internal attraction. Indeed if we estimate the temperature that would be given out by the mechanical compression of steam from a volume of 2048 to that of 1, by successively doubling the density, and supposing as above, that at each time of doubling, 25° were given out, it would be found that 12 successive operations would reduce the volume to 1, and that only 300° would be given out. But it is not right to conclude, that the same quantity of temperature would be given out at each of the successive

That an atmosphere of steam does actually surround the earth, existing independently of the other atmospheres with which however it is necessarily most intimately mixed, is I think capable of demonstration. I have endeavoured to enforce and illustrate it in several Essays in the Memoirs of the Manchester Society, and in Nicholson's Journal, to which I must refer. Now an atmosphere of any elastic fluid, whether of the weight of 30 inches of mercury, or of half an inch, must observe the same general laws; but it should seem that an atmosphere of vapour varies its temperature

condensations, though it may be nearly so for most of them: towards the conclusion, the space occupied by the solid atom or particle bears a considerable proportion to the whole space occupied by it and its atmosphere. At the first compression, the atmosphere of heat might be said to be reduced into half the space; but at the last, the reduction would be much greater, and therefore more heat given out than determined by theory.

Since writing the above, Mr. Ewart informs me that the idea respecting steam, which I had from him, is originally Mr. Watt's. In Black's Lectures, Vol. 1, page 190, the author, speaking of Mr. Watt's experiments on steam at low temperatures, observes, "we find that the latent heat of the steam is at least as much increased as the sensible heat is diminished." It is wonderful that so remarkable a fact should have been so long known and so little noticed.

less rapidly in ascending tnan the one we have of air. Something of an effect similar to what is pointed out in the 2d case above, ought therefore to be observed in our mixed atmosphere ;—namely, a condensation of vapour in the higher regions, at the same moment that evaporation is going on below.—This is actually the case almost every day, as all know from their own observation ; a cloudy stratum of air frequently exists above, whilst the region below is comparatively dry.

SECTION 9.

ON THE PHENOMENA OF THE CONGELATION OF WATER.

Several remarkable phenomena are attendant upon the congelation of water, and some of them are so different from what might be expected from analogy, that I believe no explanation according with the principles of the mechanical philosophy has been attempted, such as to account for all the appearances. This attempt is the object of the present Essay. It will be expedient previously to state the principal facts.

1. The specific gravity of ice is less than that of water in the ratio of 92 to 100.

2. When water is exposed in a large suspended jar to cool in still air of 20 or 30°, it may be cooled 2 or 3° below freezing; but if any tremulous motion take place, there appear instantly a multitude of shining hexangular *spiculæ*, floating, and slowly ascending in the water.

3. It is observed that the shoots or ramifications of ice at the commencement, and in the early stage of congelation are always at an angle of 60 or 120°.

4. Heat is given out during congelation, as much as would raise the temperature of water 150° of the new scale. The same quantity is again taken in when the ice is melted. This quantity may be $\frac{1}{40}$ of the whole heat which water of 32° contains.

5. Water is densest at 36° of the old scale, or 38° of the new: from that point it gradually *expands* by cooling or by heating alike, according to the law so often mentioned, that of the square of the temperature.

6. If water be exposed to the air, and to agitation, it cannot be cooled below 32°; the application of cold freezes a part of the water, and the mixture of ice and water requires the temperature of 32°.

7. If the water be kept still, and the cold be not severe, it may be cooled in large quantities to 25° or below, without freezing; if the water be confined in the bulb of a thermometer, it is very difficult to freeze it by any cold mixture above 15° of the old scale; but it is equally difficult to cool the water much below that temperature without its freezing. I have obtained it as low as 7 or 8°, and gradually heated it again without any part of it being frozen.

8. In the last case of what may be called *forced* cooling, the law of expansion is still observed as given above.

9. When water is cooled to 15° or below in a bulb, it retains the most perfect transparency; but if it accidentally freeze, the congelation is instantaneous, the bulb becoming in a moment opake and white like snow, and the water is projected up the stem.

10. When water is cooled below freezing, and congelation suddenly takes place, the temperature rises instantly to 32°.

In order to explain these phenomena, let it be conceived that the ultimate or smallest elements of water are all globular, and exactly of the same size; let the arrangement of these atoms be in squares, as exhibited in Fig. 1. Plate 3. so that each particle touches four others in the same horizontal plane. Conceive a second stratum of particles placed upon these in like order of squares, but so that each globule falls into the concavity of four others on the first stratum, and consequently rests upon four points, elevated 45° above the centres of the globules. A perpendicular section of such globule resting upon two diagonal globules of the square is exhibited in Fig. 3. Conceive a third stratum placed in like manner upon the second, &c. The whole being similar to a square pile of shot.—The above constitution is conceived to represent that of water at the temperature of greatest density.

To find the number of globules in a cubic vessel, the side of which is given; let $n =$ the number of particles in one line or side of the cube; then n^2 is the number in any horizontal stratum; and because a line joining the centres of two contiguous particles in different strata makes an angle of 45° with the horizontal plane, the number of strata in the given height will be $n \div$ sine of $45° = n \div \frac{1}{2}\sqrt{2}$. Whence the number of particles in the cubic vessel $= n^3 \div \frac{1}{2}\sqrt{2} = n^3\sqrt{2}$.

Now let it be supposed that the square pile is instantly drawn into the shape of a rhombus (Fig 2.); then each horizontal stratum will still consist of the same number of particles as before, only in a more condensed form, each particle being now in contact with six others. But to counteract this condensation, the several successive strata are more elevated than before, so that the pile is increased in height. A question then arises whether a vessel of given capacity will hold a greater number of particles in this or the former disposition? It must be observed, that in the last case, each particle of a superior stratum rests only on two particles of an inferior one, and is therefore elevated by the sine of 60° as represented in Fig. 4. The bases of the two piles are as $1 : \surd\frac{3}{4}$, and their heights as $\surd\frac{1}{2} : \surd\frac{3}{4}$ but the capacities are as the products of the base and height, or as $\surd\frac{1}{2} : \frac{3}{4}$; that is, as .707 to .750 nearly, or as 94 to 100. Thus it appears that the first arrangement contains more particles in a given space than the second by 6 per cent.

The last or rhomboidal arrangement is supposed to be that which the particles of water assume upon congelation. The specific gravities of ice and water should therefore be as 94 to 100. But it should be remembered that water usually contains 2 per cent. in bulk of atmospheric air: and that this air is liberated upon congelation; and is commonly entangled amongst the ice in such sort as to increase its bulk without materially increasing its weight;

this reduces the specific gravity of ice 2 per cent. or makes it 92, which agrees exactly with observation. Hence the 1st fact is explained.

The angle of a rhombus is 60°, and its supplement 120°; if any particular angles are manifested in the act of congelation, therefore we ought to expect these, agreeable to the 2d and 3d phenomena.

Whenever any remarkable change in the internal constitution of any body takes place, whether by the accession and junction of new particles, or by new arrangements of those already existing in it; some modification in the atmospheres of heat must evidently be required; though it may be difficult to estimate the quantity, and sometimes even the kind of change so produced, as in the present case. So far therefore the theory proposed agrees with the 4th phenomenon.

In order to explain the other phenomena, it will be requisite to consider more particularly the mode by which bodies are expanded by heat.—Is the expansion occasioned simply by the enlargement of the individual atmospheres of the component particles? This is the case with elastic fluids, and perhaps with solids, but certainly not with liquids. How is it possible that water should be expanded a portion represented by 1 upon the addition of a certain quantity of heat at one temperature, and by 340 upon the addition of a like quantity at another temperature, when both temperatures are remote from the absolute zero, the one perhaps 6000° and the other 6170°? The fact cannot be accounted for on any other supposition than that of a change of arrangement in the component particles; and a *gradual* change from the square to the rhomboidal arrangement is in all probability effected both by the addition and abstraction of heat. It is to be supposed then that water of the greatest possible density has its particles arranged in the square form; but if a given quantity of heat be added to, or taken from it, the particles commence their approach to the rhomboidal form, and consequently the whole is expanded, and that the same by the same change of temperature, whether above or below that point.

If heat be taken away from water of 38°, then expansion is the consequence, and a moderate inclination of the particles towards the rhomboidal form; but this only extends a small way whilst the mass is subject to a tremulous motion, so as to relieve the obstructions occasioned by friction; by the energy of certain affinities, the new form is completed in moment, and a portion of ice formed; heat is then given out which retards the subsequent formation, till at last the whole is congealed. This is the ordinary process of congelation. But if the mass of water cooled is kept in a state of perfect tranquillity, the gradual approach to the rhomboidal form can be carried much farther; the expansion goes on according to the usual manner, and the slight friction or adhesion of the particles is sufficient to counteract the balance of energies in favour of the new formation, till some accidental tremor contributes to adjust the equilibrium. A similar operation is performed when we lay a piece of iron on a table, and hold a magnet gradually nearer and nearer; the proximity of the approach, without contact, is much assisted by guarding against any tremulous motion of the table. Hence the rest of the phenomena are accounted for.

CHAP. II.

ON THE

CONSTITUTION OF BODIES.

THERE are three distinctions in the kinds of bodies, or three states, which have more especially claimed the attention of philosophical chemists; namely, those which are marked by the terms *elastic fluids, liquids, and solids.* A very familiar instance is exhibited to us in water, of a body, which, in certain circumstances, is capable of assuming all the three states. In steam we recognise a perfectly elastic fluid, in water, a perfect liquid, and in ice a complete solid. These observations have tacitly led to the conclusion which seems universally adopted, that all bodies of sensible magnitude, whether liquid or solid, are constituted of a vast number of extremely small particles, or atoms of matter bound together by a force of attraction, which is more or less powerful according to circumstances, and which as it endeavours to prevent their separation, is very properly called in that view, *attraction of cohesion*; but as it collects them from a dispersed state (as from steam into water) it is called, *attraction of aggregation*, or more simply, *affinity*. Whatever names it may go by, they still signify one and the same power. It is not my design to call in question this conclusion, which appears completely satisfactory; but to shew that we have hitherto made no use of it, and that the consequence of the neglect, has been a very obscure view of chemical agency, which is daily growing more so in proportion to the new lights attempted to be thrown upon it.

The opinions I more particularly allude to, are those of Berthollet on the Laws of chemical affinity; such as that chemical agency is proportional to the mass, and that in all chemical unions, there exist insensible gradations in the proportions of the constituent principles. The inconsistence of these opinions, both with reason and observation, cannot, I think, fail to strike every one who takes a proper view of the phenomena.

Whether the ultimate particles of a body, such as water, are all alike, that is, of the same figure, weight, &c. is a question of some importance. From what is known, we have no reason to apprehend a diversity in these particulars: if it does exist in water, it must equally exist in the elements constituting water, namely, hydrogen and oxygen. Now it is scarcely possible to conceive how the aggregates of dissimilar particles should be so uniformly the same. If some of the particles of water were heavier than others, if a parcel of the liquid on any occasion were constituted principally of these heavier particles, it must be supposed to affect the specific gravity of the mass, a circumstance not known. Similar observations may be made on other substances. Therefore we may conclude that *the ultimate particles of all homogeneous bodies are perfectly alike in weight, figure, &c.* In other words, every particle of water is like every other particle of water; every particle of hydrogen is like every other particle of hydrogen, &c.

Besides the force of attraction, which, in one character or another, belongs universally to ponderable bodies, we find another force that is likewise universal, or acts upon all matter which comes under our cognisance, namely, a force of repulsion. This is now generally, and I think properly, ascribed to the agency of heat. An atmosphere of this subtile fluid constantly surrounds the atoms of all bodies, and prevents them from being drawn into actual contact. This appears to be satisfactorily proved by the observation, that the bulk of a body may be diminished by abstracting some of its heat: But from what has been stated in the last section, it should seem that enlargement and diminution of bulk depend perhaps more on the arrangement, than on the size of the ultimate particles. Be this as it may, we cannot avoid inferring from the preceding doctrine on heat, and particularly from the section on the natural zero of temperature, that solid bodies, such as ice, contain a large portion, perhaps $\frac{4}{5}$ of the heat which the same are found to contain in an elastic state, as steam.

We are now to consider how these two great antagonist powers of attraction and repulsion are adjusted, so as to allow of the three different states of *elastic fluids, liquids, and solids*. We shall divide the subject into four Sections; namely, first, *on the constitution of pure elastic fluids;* second, *on the constitution of mixed elastic fluids;* third, *on the constitution of liquids*, and fourth, *on the constitution of solids*.

SECTION 1.

ON THE CONSTITUTION OF PURE ELASTIC FLUIDS.

A pure elastic fluid is one, the constituent particles of which are all alike, or in no way distinguishable. Steam, or aqueous vapour, hydrogenous gas, oxygenous gas, azotic gas,* and several others are of this kind. These fluids are constituted of particles possessing very diffuse atmospheres of heat, the capacity or bulk of the atmosphere being often one or two thousand times that of the particle in a liquid or solid form. Whatever therefore may be the shape or figure of the solid atom abstractedly, when surrounded by such an atmosphere it must be globular; but as all the globules in any small given volume are subject to the same pressure, they must be equal in bulk, and will therefore be arranged in horizontal strata, like a pile of shot. A volume

* The novice will all along understand that several chemical subjects are necessarily introduced before their general history and character can be discussed.

of elastic fluid is found to expand whenever the pressure is taken off. This proves that the repulsion exceeds the attraction in such case. The absolute attraction and repulsion of the particles of an elastic fluid, we have no means of estimating, though we can have little doubt but that the cotemporary energy of both is great; but the excess of the repulsive energy above the attractive can be estimated, and the law of increase and diminution be ascertained in many cases. Thus in steam, the density may be taken at $\frac{1}{1728}$ that of water; consequently each particle of steam has 12 times the diameter that one of water has, and must press upon 144 particles of a watery surface; but the pressure upon each is equivalent to that of a column of water of 34 feet; therefore the excess of the elastic force in a particle of steam is equal to the weight of a column of particles of water, whose height is $34 \times 144 = 4896$ feet. And further, this elastic force decreases as the distance of the particles increases. With respect to steam and other elastic fluids then, the force of cohesion is entirely counteracted by that of repulsion, and the only force which is efficacious to move the particles is the excess of the repulsion above the attraction. Thus, if the attraction be as 10 and the repulsion as 12, the effective repulsive force is as 2. It appears then, that an elastic fluid, so far from requiring any force to separate its particles, it always requires a force to retain them in their situation, or to prevent their separation.

A vessel full of any pure elastic fluid presents to the imagination a picture like one full of small shot. The globules are all of the same size; but the particles of the fluid differ from those of the shot, in that they are constituted of an exceedingly small central atom of solid matter, which is surrounded by an atmosphere of heat, of great density next the atom, but gradually growing rarer according to some power of the distance; whereas those of the shot are globules, uniformly hard throughout, and surrounded with atmospheres of heat of no comparative magnitude.

It is known from experience, that the force of a mass of elastic fluid is directly as the density. Whence is derived the law already mentioned, that the repulsive power of each particle is inversely as its diameter. That is, the *apparent* repulsive power, if we may so speak; for the real or absolute force of repulsion is not known, as long as we remain ignorant of the attractive force. When we expand any volume of elastic fluid, its particles are enlarged, without any material change in the quantity of their heat; it follows then, that the density of the atmospheres of heat must fluctuate with the pressure. Thus, suppose a measure of air were expanded into 8 measures; then, because the diameters of the elastic particles are as the cube root of the space, the distances of the particles would be twice as great as before, and the elastic atmospheres would occupy nearly 8 times the space they did before, with nearly the same quantity of heat: whence we see that these atmospheres must be diminished in density in nearly the same ratio as the mass of elastic fluid.

Some elastic fluids, as hydrogen, oxygen, &c. resist any pressure that has yet been applied to them. In such then it is evident the repulsive force of heat is more than a match for the affinity of the particles, and the external pressure united. To what extent this would continue we cannot say; but from analogy we might apprehend that a still greater pressure would succeed in giving the attractive force the superiority, when the elastic fluid would become a liquid or solid. In other elastic fluids, as steam, upon the application of compression to a certain degree, the elasticity apparently ceases altogether, and the particles collect in small drops of liquid, and fall down. This phenomenon requires explanation.

From the very abrupt transition of steam

from a volume of 1700 to that of 1, without any material increase of pressure, one would be inclined to think that the condensation of it was owing to the *breaking* of a spring, rather than to the *curbing* of one. The last however I believe is the fact. The condensation arises from the action of affinity becoming superior to that of heat, by which the latter is overruled, but not weakened. As the approximation of the particles takes place, their repulsion increases from the condensation of the heat, but their affinity increases, it should seem, in a still greater ratio, till the approximation has attained a certain degree, when an equilibrium between those two powers takes place, and the liquid, water, is the result. That this is the true explanation we may learn from what has been stated at page 131; wherein it is shewn that the heat given off by the condensation of steam, is in all probability no more than would be given off by any permanently elastic fluid, could it be mechanically condensed into the like volume, and is moreover a small portion of the whole heat previously in combination. As far then as the heat is concerned in this phenomenon, the circumstances would be the same, whether the approximation of the particles was the effect of affinity, or of external mechanical force.

The constitution of a liquid, as water, must then be conceived to be that of an aggregate of particles, exercising in a most powerful manner the forces of attraction and repulsion, but nearly in an equal degree.—Of this more in the sequel.

SECTION 2.

ON THE CONSTITUTION OF MIXED ELASTIC FLUIDS.

When two or more elastic fluids, whose particles do not unite chemically upon mixture, are brought together, one measure of each, they occupy the space of two measures, but become uniformly diffused through each other, and remain so, whatever may be their specific gravities. The fact admits of no doubt; but explanations have been given in various ways, and none of them completely satisfactory. As the subject is one of primary importance in forming a system of chemical principles, we must enter somewhat more fully into the discussion.

Dr. Priestley was one of the earliest to notice the fact: it naturally struck him with surprise, that two elastic fluids, having apparently no affinity for each other, should not arrange themselves according to their specific gravities, as liquids do in like circumstances. Though he found this was not the case after the elastic fluids had once been thoroughly mixed, yet he suggests it as probable, that if two of such fluids could be exposed to each other without agitation, the one specifically heavier would retain its lower situation. He does not so much as hint at such gases being retained in a mixed state by affinity. With regard to his suggestion of two gases being carefully exposed to each other without agitation, I made a series of experiments expressly to determine the question, the results of which are given in the Manch. Memoirs, Vol. 1. *new series.* From these it seems to be decided that gases always intermingle and gradually diffuse themselves amongst each other, if exposed ever so carefully; but it requires a considerable time to produce a complete intermixture, when the surface of communication is small. This time may vary from a minute, to a day or more, according to the quantity of the gases and the freedom of communication.

When or by whom the notion of mixed gases being held together by chemical affinity was first propagated, I do not know; but it seems probable that the notion of water being dissolved in air, led to that of air being dissolved in air.—Philosophers found that water gradually disappeared or evaporated in air, and increased its elasticity; but steam at a low temperature was known to be unable to overcome the resistance of the air, therefore the agency of affinity was necessary to account for the effect. In the permanently elastic fluids indeed, this agency did not seem to be so much wanted, as they are all able to support themselves; but the diffusion through each other was a circumstance which did not admit of an easy solution any other way. In regard to the solution of water in air, it was natural to suppose, nay, one might almost have been satisfied without the aid of experiment, that the different gases would have had different affinities for water, and that the quantities of water dissolved in like circumstances, would have varied according to the nature of the gas. Saussure found however that there was no difference in this respect in the solvent powers of carbonic acid, hydrogen gas, and common air.—It might be expected that at least the *density* of the gas would have some influence upon its solvent powers, that air of half density would take half the water, or the quantity of water would diminish in some proportion to the density; but even here again we are disappointed; whatever be the rarefaction, if water be present, the vapour produces the same elasticity, and the hygrometer finally settles at extreme moisture, as in air of common density in like circumstances. These facts are sufficient to create extreme difficulty in the conception how any principle of affinity or *cohesion* between air and water can be the agent. It is truly astonishing that the same quantity of vapour should cohere to *one* particle of air in a given space, as to *one thousand* in the same space. But the wonder does not cease here; a torricellian vacuum dissolves water; and in this instance we have vapour existing independently of air at all temperatures; what makes it still more remarkable is, the vapour in such vacuum is precisely the same in quantity and force as in the like volume of any kind of air of extreme moisture.

These and other considerations which occurred to me some years ago, were sufficient to make me altogether abandon the hypothesis of air dissolving water, and to explain the phenomena some other way, or to acknowledge they were inexplicable. In the autumn of 1801, I hit upon an idea which seemed to be exactly calculated to explain the phenomena of vapour; it gave rise to a great variety of experiments upon which a series of essays were founded, which were read before the Literary and Philosophical Society of Manchester, and published in the 5th Vol. of their memoirs, 1802.

The distinguishing feature of the new theory was, that the particles of one gas are not elastic or repulsive in regard to the particles of another gas, but only to the particles of their own kind. Consequently when a vessel contains a mixture of two such elastic fluids, each acts independently upon the vessel, with its proper elasticity just as if the other were absent, whilst no mutual action between the fluids themselves is observed. This position most effectually provided for the existence of vapour of any temperature in the atmosphere, because it could have nothing but its own weight to support; and it was perfectly obvious why neither more nor less vapour could exist in air of extreme moisture, than in a vacuum of the same temperature. So far then the great object of the theory was attained. The law of the condensation of vapour in the atmosphere by cold, was evidently the same on this scheme, as that of the condensation of pure steam, and experience was found to confirm the conclusion at all temperatures. The only thing now wanting to completely establish the independent existence of aqueous vapour in the atmosphere, was the conformity of other liquids to water, in regard to the diffusion and condensation of their vapour. This was found to take place in several liquids, and particularly in sulphuric ether, one which was most likely to shew any anomaly to advantage if it existed, on account of the great change of expansibility in its vapour at ordinary temperatures. The existence of vapour in the atmosphere and its occasional condensation were thus accounted for; but another question remained, how does it rise from a surface of water subject to the pressure of the atmosphere? The consideration of this made no part of the essays abovementioned, it being apprehended, that if the other two points could be obtained by any theory, this third too, would, in the sequel, be accomplished.

From the novelty, both in the theory and the experiments, and their importance, provided they were correct, the essays were soon circulated, both at home and abroad. The new facts and experiments were highly valued, some of the latter were repeated, and found correct, and none of the results, as far as I know, have been controverted; but the theory was almost universally misunderstood, and consequently reprobated. This must have have arisen partly at least from my being too concise, and not sufficiently clear in its exposition.

Dr. Thomson was the first, as far as I know, who publicly animadverted upon the theory; this gentleman, so well known for his excellent System of Chemistry, observed in the first edition of that work, that the theory would not account for the equal distribution of gases; but that, granting the supposition of one gas neither attracting nor repelling another, the two must still arrange themselves according to their specific gravity. But the most general objection to it was quite of a different kind; it was admitted, that the theory was adapted so as to obtain the most uniform and permanent diffusion of gases; but it was urged, that as one gas was as a vacuum to another, a measure of any gas being put to a measure of another, the two measures ought to occupy the space of one measure only. Finding that my views on the subject were thus misapprehended, I wrote an illustration of the theory, which was published in the 3d Vol. of Nicholson's Journal, for November, 1802. In that paper I endeavoured to point out the conditions of mixed gases more at large, according to my hypothesis; and particularly touched upon the discriminating feature of it, that of two particles of any gas A, repelling each other by the known stated law, whilst one or more particles of another gas B, were interposed in a direct line, without at all affecting the reciprocal action of the said two particles of A. Or, if any particle of B were casually to come in contact with one of A, and press against it, this pressure did not preclude the cotemporary action of all the surrounding particles of A upon the one in contact with B. In this respect the mutual action of particles of the same gas was represented as resembling magnetic action, which is not disturbed by the intervention of a body not magnetic.

As the subject has since received the animadversions of several authors, which it is expedient to notice more or less, it will be proper to point out the order intended to be pursued. First, I shall consider the objections to the new theory made by the several authors, with their own views on the subject; and then shall give what modifications of the theory, the experience and reflection of succeeding time have suggested to me. The authors are Berthollet, Dr. Thomson, Mr. Murray, Dr. Henry, and Mr. Gough.

Berthollet in his Chemical Statics (1804) has given a chapter on the constitution of the atmosphere, in which he has entered largely into a discussion of the new theory. This celebrated chemist, upon comparing the results of experiments made by De Luc, Saussure, Volta, Lavoisier, Watt, &c. together with those of Gay Lussac, and his own, gives his full assent to the fact, that vapours of every kind increase the elasticity of each species of gas alike, and just as much as the force of the said vapours in vacuo; and not only so, but that the specific gravity of vapour in air and vapour in vacuo is in all cases the same (Vol. 1. Sect. 4.) Consequently he adopts the theorem for finding the quantity of vapour which a given volume of air can dissolve, which I have laid down; namely,

$$s = \frac{p}{p-f}$$

where p represents the pressure upon a given volume (1) of dry air, expressed in inches of mercury, $f =$ the force of the vapour in vacuo at the temperature, in inches of mercury, and $s =$ the space which the mixture of air and vapour occupies under the given pressure, p, after saturation. So far therefore we perfectly agree: but he objects to the theory by which I attempt to explain these phenomena, and substitutes another of his own.

The first objection I shall notice is one that

clearly shews Berthollet either does not understand, or does not rightly apply the theory he opposes; he says, "If one gas occupied the interstices of another, as though they were vacancies, there would not be any augmentation of volume when aqueous or ethereal vapour was combined with the air; nevertheless there is one proportional to the quantity of vapour added: humidity should increase the specific gravity of the air, whereas it renders it specificallv lighter, as has been already noticed by Newton." This is the objection which has been so frequently urged; it has even been stated by Mr. Gough, if I understand him aright, in almost the same words (Nicholson's Journal, Vol. 9, page 162); yet this last gentleman is profoundly skilled in the mechanical action of fluids. Let a tall cylindrical glass vessel containing dry air be inverted over mercury, and a portion of the air drawn out by a syphon, till an equilibrium of pressure is established within and without; let a small portion of water, ether, &c. be then thrown up into the vessel; the vapour rises and occupies the interstices of the air as a void; but what is the obvious consequence? Why, the surface of the mercury being now pressed both by the dry air, and by the new raised vapour, is more pressed within than without, and an enlargement of the volume of air is unavoidable, in order to restore the equilibrium. Again in the open air: suppose there were no aqueous atmosphere around the earth, only an azotic one = 23 inches of mercury, and an oxygenous one = 6 inches. The air being thus perfectly dry, evaporation would commence with great speed. The vapour first formed being constantly urged to ascend by that below, and as constantly resisted by the air, must, in the first instance, dilate the other two atmospheres; (for, the ascending steam adds its force to the upward elasticity of the two gases, and in part alleviates their pressure, the necessary consequence of which is dilatation.) At last when all the vapour has ascended, that the temperature will admit of, the aqueous atmosphere attains an equilibrium; it no longer presses upon the other two, but upon the earth; the others return to their original density and pressure throughout. In this case it is true, there would not be any augmentation of volume when aqueous vapour was combined with the air; humidity would increase the weight of the congregated atmospheres, but diminish their specific gravity under a given pressure. One would have thought that this solution of the phenomenon upon my hypothesis was too obvious to escape the notice of any one in any degree conversant with pneumatic chemistry. Berthollet indeed enquires, "Is such a divsion of the same pressure of the atmosphere analogous with any physical property yet known? Can it be conceived that an elastic substance exists, which adds its volume to that of another, and which nevertheless does not act on it by its expansive force?" Certainly; we can not only conceive it, but bring an instance that must be allowed to be in point. Two magnets repel each other, that is, act upon each other with an expansive force, yet they do not act upon other bodies in the same way, but merely as inelastic bodies; and this no doubt would be the same if they were reduced to atoms: So two particles of the same kind of air may act upon each other elastically, and upon other bodies inelastically, and therefore not at all, unless when in contact.

Berthollet observes, "Hydrogen gas and oxygen gas form water in a given circumstance; azotic gas, and oxygen gas, can also produce nitric acid; but the reciprocal action which decides the combinations cannot be considered as a force commencing at the precise moment at which it is manifested, it must have existed long before producing its effect, and increases gradually till it becomes preponderant." It is no doubt true that the opposite powers of attraction and repulsion are frequently, perhaps constantly, energetic at the same instant; but the effect produced in those cases arises from the difference of the two powers. When the excess of the repulsive power above the attractive in different gases is comparatively small and insignificant, it constitutes that character which may be denominated neutral, and which I supposed to exist in the class of mixed gases which are not observed to manifest any sign of chemical union. I would not be understood to deny an energetic affinity between oxygen and hydrogen, &c. in a mixed state; but that affinity is more than counterbalanced by the repulsion of the heat, except in circumstances which it is not necessary at present to consider.

Again, " Azotic gas comports itself with oxygen gas, in the changes occasioned by temperature and pressure, precisely like one and the same gas: Is it necessary to have recourse to a supposition which obliges us to admit so great a difference of action without an ostensible cause?" It is possible this may appear an objection to a person who does not understand the theory, but it never can be any to one who does. If a mixture of gas, such as atmospheric air, containing azote pressing with a force equal to 24 inches of mercury, and oxygen with a force equal to 6 inches, were suddenly condensed into half the compass, the azotic gas would then evidently, on my hypothesis, press with a force = 48 inches, and the oxygen with a force = 12 inches, making together 60 inches, just the same as any simple gas. And a similar change in the elasticity of each would take place by heat and cold. Will the opposite theory of Berthollet be equally free from this objection? We shall presently examine it.

Another objection is derived from the very considerable time requisite for a body of hydrogen to descend into one of carbonic acid; if one gas were as a vacuum for another, why is the equilibrium not instantly established? This objection is certainly plausible; we shall consider it more at large hereafter.

In speaking of the pressure of the atmosphere retaining water in a liquid state, which I deny, Berthollet adopts the idea of Lavoisier, "that without it the moleculæ would be infinitely dispersed, and that nothing would limit their separation, unless their own weight should collect them to form an atmosphere." This, I may remark, is not the language dictated by a correct notion on the subject. Suppose our atmosphere were annihilated, and the waters on the surface of the globe were instantly expanded into steam; surely the action of gravity would collect the moleculæ into an atmosphere of similar constitution to the one we now possess; but suppose the whole mass of water evaporated amounted in weight to 30 inches of mercury, how could it support its own weight at the common temperature? It would in a short time be condensed into water merely by its weight, leaving a small portion, such as the temperature could support, amounting perhaps to half an inch of mercury in weight, as a permanent atmosphere, which would effectually prevent any more vapour from rising, unless there were an increase of temperature. Does not every one know that water and other liquids can exist in a Torricellian vacuum at low temperatures solely by the pressure of vapour arising from them? What need then of the pressure of the atmosphere in order to prevent an excess of vapourisation?

After having concluded that "without the pressure of the ærial atmosphere, liquids would pass to the elastic state," Berthollet proceeds in the very next paragraph to shew that the quantity of vapour in the atmosphere may in fact be much *more* than would exist if the atmosphere were suppressed, and hence infers, "that the variations of the barometer occasioned by those of the humidity of the atmosphere may be much greater than was believed by Saussure and Deluc." I cannot see how the author reconciles the opposite conclusions.

The experiments of Fontana on the distillation of water and ether in close vessels containing air, are adduced to prove, that vapours do not penetrate air without resistance. This is true no doubt; vapour cannot make its way in such circumstances through a long and circuitous route without time, and if the external atmosphere keep the vessel cool, the vapour may be condensed by its sides, and fall down in a liquid form as fast as it is generated, without ever penetrating in any sensible quantity to its remote extremity.

We come now to the consideration of that theory which Berthollet adopts in his explanation of the phenomena of gaseous mixtures. According to his theory, there are two degrees of affinity. The one is strong, makes the particles of bodies approach nearer to each other, and generally expels heat: the effect of this may be called *combination*; for instance, when oxygen gas is put to nitrous gas, the two combine, give out heat, are condensed in volume, and become possessed of properties different from what they had previously. The other is weak; it does not sensibly condense the volume of any mixture, nor give out heat, nor change the properties of the ingredients; its effect may be called *solution* or *dissolution*; for instance, when oxygen gas and azotic gas are mixed in due proportion, they constitute atmospheric air, in which they retain their distinguishing properties.

It is upon this supposed *solution* of one elastic fluid in another that I intend to make a few observations. That I have not misrepresented the author's ideas, will, I think, appear from the following quotations. "When different gases are mixed, whose action is confined to this solution, no change is observed in the temperature, or in the volume resulting from the mixture; hence it may be concluded, that this mutual action of two gases does not produce any condensation, and that it cannot surmount the effort of the elasticity, or the affinity for caloric, so that the properties of each gas are not sensibly changed—." "Although both the solution and combination of two gases are the effect of a chemical action, which only differs in its intensity, a real difference may be established between them, because there is a very material difference between the results: the combination of two gases always leads to a condensation of their volume, and gives rise to new properties; on their solution, the gases share in common the changes arising from compression and temperature, and preserve their individual properties, which are only diminished in the ratio of the slight action which holds them united." (Page 198.) "The mutual affinity of the gases can, therefore, produce between them an effect which is greater than their difference of specific gravity, but which is inferior to the elastic tension which belongs to each molecule of both, so that the volume is not changed by this action; the liquids which take the elastic state, comport themselves afterwards like the gases." (Page 218.) "Solution must be distinguished from combination, not only because in the first, each of the substances is retained by an affinity so weak, that it preserves its dimensions.—" (Page 219.) Again, "It cannot be doubted, that the parts of elastic fluids are *not* endued with the force of cohesion, as the substances dissolved by them undergo an equal distribution, which could not happen but by the means of a reciprocal chemical attraction; that which constitutes the force of cohesion." (Researches into the Laws of chemical affinity, Eng. Trans. page 57.) Here the translator has, I apprehend, mistaken the English idiom. The author means to say, that the parts of elastic fluids *are* endued with the force of cohesion; but this he applies only to heterogeneous particles. He certainly does not mean that the particles of homogeneous elastic fluids possess the force of cohesion.

Newton has demonstrated from the phenomena of condensation and rarefaction that elastic fluids are constituted of particles, which repel one another by forces which increase in proportion as the distance of their centres diminishes: in other words, the forces are reciprocally as the distances. This deduction will stand as long as the Laws of elastic fluids continue to be what they are. What a pity it is that all who attempt to reason, or to theorise respecting the constitution of elastic fluids, should not make themselves thoroughly acquainted with this immutable Law, and constantly hold it in their view whenever they start any new project! When we contemplate a mixture of oxygenous and hydrogenous gas, what does Berthollet conceive, are the particles that repel each other according to the Newtonian Law? The mixture *must* consist of such; and he ought in the very first instance to have informed us what constitutes the *unity* of a particle in his solution. If he grants that each particle of oxygen retains its unity, and each particle of hydrogen does the same, then we must conclude that the mutual action of two particles of oxygen is the same as that of a particle of oxygen, and one of hydrogen, namely, a repulsion according to the Law above stated, which effectually destroys the supposed solution by chemical agency. But if it be supposed that each particle of hydrogen attaches itself to a particle of oxygen, and the two particles so united form *one*, from which the repulsive energy emanates; then the new elastic fluid may perfectly conform to the Newtonian Law; in this case a true saturation will take place when the number of particles of hydrogen and oxygen in a mixture happen to be equal, or at least in the ratio of some simple numbers, such as 1 to 2, 1 to 3, &c. Now something like this does actually take place when a real combination is formed, as for instance, steam, and nitric acid formed of a mixture of oxygen and nitrous gas. Here we have new elastic fluids, the atoms of which repel one another by the common Law, heat is given out, a great condensation of volume ensues, and the new fluids differ from their constituents in their chemical relations. It remains then to determine whether, in the instance of solution, all these effects take place in a "slight" degree; that is, in so small a degree as not to be cognisable to any of the senses. It certainly requires an extraordinary stretch of the imagination to admit the affirmative.

One great reason for the adoption of this, or any other theory on the subject, arises from the phenomena of the evaporation of water. How is water taken up and retained in the atmosphere? It cannot be in the state of vapour, it is said, because the pressure is too great: there must therefore be a true chemical solution. But when we consider that the surface of water is subject to a pressure equal to 30 inches of mercury, and besides this pressure, there is a *sensible* affinity between the particles of water themselves; how does the *insensible* affinity of the atmosphere for water overcome both these powers? It is to me quite inexplicable upon this hypothesis, the leading object of which is to account for this very phenomenon. Further, if a particle of air has attached a particle of water to it, what reason can be assigned why a superior particle of air should rob an inferior one of its property, when each particle possesses the same power? If a portion of common salt be dissolved in water and a little muriatic acid added; is there any reason to suppose the additional acid displaces that already combined with the soda, and that upon evaporation the salt is not obtained with the identical acid it previously had? Or, if oxygen gas be confined by water, is there any reason to suppose that the hydrogen of the water is constantly giving its oxygen to the air and receiving an equal quantity from the same? Perhaps it will be said in the case of air dissolving water, that it is not the affection of one particle for one, it is that of a mass of particles for another mass; it is the united action of all the atoms in the atmosphere upon the water, which raises up a particle. But as all these energies are reciprocal, the water must have a like action on the air, and then an atmosphere over water would press downward by a force greater than its weight, which is contradicted by experience.

When two measures of hydrogen and one of oxygen gas are mixed, and fired by the electric spark, the whole is converted into steam, and if the pressure be great, this steam becomes water. It is most probable then that there is the same number of particles in two measures of hydrogen as in one of oxygen. Suppose then three measures of hydrogen are mixed with one of oxygen, and this slight affinity operates as usual; how is the union effected? According to the principle of equal division, each atom of oxygen ought to have *one atom and a half* of hydrogen attached to it; but this is impossible; one half of the atoms of oxygen must then take two of hydrogen, and the other half, one each. But the former would be specifically lighter than the latter, and ought to be found at the top of the solution; nothing like this is however observed on any occasion.

Much more might be advanced to shew the absurdity of this doctrine of the solution of one gas in another, and the insufficiency of it to explain any of the phenomena; indeed I should not have dwelt so long upon it, had I not apprehended that respectable authority was likely to give it credit, more than any arguments in its behalf derived from physical principles.

Dr. Thomson, in the 3d Edition of his System of Chemistry, has entered into a discussion on the subject of mixed gases; he seems to comprehend the excellence and defects of my notions on these subjects, with great acuteness. He does not conclude with Berthollet, that on my hypothesis, "there would not be any augmentation of volume when aqueous and ethereal vapour was combined with the air," which has been so common an objection. There is however one objection which this gentleman urges, that shews he does not completely understand the mechanism of my hypothesis. At page 448, Vol. 3. he observes that from the principles of hydrostatics, " each particle of a fluid sustains the whole pressure. Nor can I perceive any reason why this principle should not hold, even on the supposition that Dalton's hypothesis is well founded." Upon this I would observe, that when once an equilibrium is established in any mixture of gases, each particle of gas is pressed as if by the surrounding particles *of its own kind only.* It is in the renunciation of that hydrostatical principle that the leading feature of the theory consists. The lowest particle of oxygen in the atmosphere sustains the weight of all the particles of oxygen above it, and the weight of no other. It was therefore a maxim with me, that every particle of gas is equally pressed in every direction, but the pressure arises from the particles of its own kind only. Indeed when a measure of oxygen is put to a measure of azote, at the moment the two surfaces come in contact, the particles of each gas press against those of the other with their full force; but the two gases get gradually intermingled, and the force which each particle has to sustain proportionally diminishes, till at last it becomes the same as that of the original gas dilated to twice its volume. The ratio of the forces is as the cube root of the spaces inversely; that is, as $\sqrt[3]{2} : 1$, or as 1.26 : 1 nearly. In such a mixture as has just been mentioned, then, the common hypothesis supposes the pressure of each particle of gas to be 1.26; whereas mine supposes it only to be 1; but the sum of the pressure of both gases on the containing vessel, or any other surface, is exactly the same on both hypotheses.

Excepting the above objection, all the rest which Dr. Thomson has made, are of a nature not so easily to be obviated;—he takes notice of the considerable time which elapses before two gases are completely diffused through each other, as Berthollet has done, and conceives this fact, makes against the supposition, that one gas is as a vacuum to another. He further objects, that if the particles of different gases are inelastic to each other; then a particle of oxygen coming into actual contact with one of hydrogen, ought to unite with it, and form a particle of water; but, on the other hand, he properly observes, that the great facility with which such combinations are effected in such instances as a mixture of nitrous and oxygen gas, is an argument in favour of the hypothesis.- -Dr. Thomson founds another objection upon the facility of certain combinations, when one of the ingredients is in a *nascent* form;

that is, just upon the point of assuming the elastic state; this, he observes, "seems incompatible with the hypothesis, that gases are not mutually elastic." Upon the whole, Dr. Thomson inclines to the opinion of Berthollet, that gases have the property of dissolving each other; and admits, "however problematical it may appear at first view, that the gases not only mutually repel each other, but likewise mutually attract." I have no doubt if he had taken due time to consider this conclusion, he would, with me, have pronounced it absurd: but of this again in the sequel.

With regard to the objection, that one gas makes a more durable resistance to the entrance of another, than it ought to do on my hypothesis: This occurred to me in a very early period of my speculations; I devised the train of reasoning which appeared to obviate the objection; but it being necessarily of a mathematical nature, I did not wish to obtrude it upon the notice of chemical philosophers, but rather to wait till it was called for.—The resistance which any medium makes to the motion of a body, depends upon the surface of that body, and is greater as the surface is greater, all other circumstances being the same. A ball of lead 1 inch in diameter meets with a certain resistance in falling through the air; but the same ball, being made into a thousand smaller ones of $\frac{1}{10}$ of an inch diameter, and falling with the same velocity, meets with 10 times the resistance it did before: because the force of gravity increases as the *cube* of the diameter of any particle, and the resistance only as the *square* of the diameter. Hence it appears, that in order to increase the resistance of particles moving in any medium, it is only necessary to divide them, and that the resistance will be a maximum when the division is a maximum. We have only then to consider particles of lead falling through air by their own gravity, and we may have an idea of the resistance of one gas entering another, *only the particles of lead must be conceived to be infinitely small*, if I may be allowed the expression. Here we shall find great resistance, and yet no one, I should suppose, will say, that the air and the lead are mutually elastic.

The other two objections of Dr. Thomson, I shall wave the consideration of at present.

Mr. Murray has lately edited a system of chemistry, in which he has given a very clear description of the phenomena of the atmosphere, and of other similar mixtures of elastic fluids. He has ably discussed the different theories that have been proposed on the subject, and given a perspicuous view of mine, which he thinks is ingenious, and calculated to explain several of the phenomena well, but upon the whole, not equally satisfactory with that which he adopts. He does not object to the mechanism of my hypothesis in regard to the independent elasticity of the several gases entering into any mixture, but argues that the phenomena do not require so extraordinary a postulatum ; and more particularly disapproves of the application of my theory to account for evaporation.

The principal feature in Mr. Murray's theory, and which he thinks distinguishes it from mine, is "that between mixed gases, which are capable, under any circumstances of combining, an attraction must always be exerted." It is unnecessary to recount the arguments on behalf of this conclusion, because it will not be controverted. Mr. Murray announces his views of the constitution of the atmosphere, as follows: "Perhaps that chemical attraction which subsists between the solid bases of these gases, but which, when they are merely mixed together, cannot, from the distance at which their particles are placed by the repulsive power of caloric, bring them into intimate union, may still be so far exerted, as to prevent their separation: or, they may be retained in mixture by that force of adhesion, which, exerted at the surfaces of many bodies, retains them in contact with considerable force." He supports these notions at length by various observations, and repeats some of the observations of Berthollet, whose doctrine on this subject, as has been seen, is nearly the same.

Before we animadvert on these principles, it may be convenient to extend the first a little farther, and to adopt as a maxim, "that between the particles of *pure* gases, which are capable under any circumstances of combining, an attraction must always be exerted." This, Mr. Murray cannot certainly object to, in the case of steam, a pure elastic fluid, the particles of which are known in certain circumstances to combine. Nor will it be said that steam and a permanent gas are different; for he justly observes, "this distinction (between gases and vapours) is merely relative, and arises from the difference of temperature at which they are formed; the state with regard to each, while they exist in it, is precisely the same." Is steam then constituted of particles in which the attraction is so far exerted as to prevent their separation? No: they exhibit no traces of attraction, more than the like number of particles of oxygen do, when in the gaseous form. What then is the conclusion? It is this: *notwithstanding it must be allowed, that all bodies, at all times, and in every situation, attract one another; yet in certain circumstances, they are likewise actuated by a repulsive power; the only efficient motive force is then the difference of these two powers.*

From the circumstance of gases mixing together without experiencing any sensible diminution of volume, the advocates for the agency of chemical affinity, characterise it as a "slight action," and "a weak reciprocal action:" So far I think they are consistent; but when we hear of this affinity being so far exerted as to prevent the separation of elastic particles, I do not conceive with what propriety it can be called weak. Suppose this affinity should be exercised in the case of steam of 212°; then the attraction becoming equal to the repulsion, the force which any one particle would exercise must be equal to the weight of a column of water of 4896 feet high. (See page 146.)

It is somewhat remarkable that those gases which are known to combine occasionally, as azote and oxygen, and those which are never known to combine, as hydrogen and carbonic acid, should dissolve one another with equal facility; nay, these last exercise this solvent power with more effect than the former; for, hydrogen can draw up carbonic acid from the bottom to the top of any vessel, notwithstanding the latter is 20 times the specific gravity of the former. One would have thought that a force of adhesion was more to be expected in the particles of steam, than in a mixture of hydrogen and carbonic acid. But it is the business of those who adopt the theory of the mutual solution of gases to explain these difficulties.

In a mixture where are 8 particles of oxygen for 1 of hydrogen, it is demonstrable that the central distances of the particles of hydrogen are at a medium twice as great as those of oxygen. Now supposing the central distance of two adjacent particles of hydrogen to be denoted by 12, query, what is supposed to be the central distance of any one particle of hydrogen from that one particle, or those particles of oxygen with which it is connected by this weak chemical union? It would be well if those who understand and maintain the doctrine of chemical solution would represent how they conceive this to be; it would enable those who are desirous to learn, to obtain a clear idea of the system, and those who are dissatisfied with it, to point out its defects with more precision. The greatest possible central distance would be $8\frac{1}{2}$ in the above instance, and the least might perhaps be 1. Berthollet, who decries the diagram by which I endeavoured to illustrate my ideas on thi subject, has not given us any precise information, either verbally or otherwise, relative to the collocation of the heterogeneous particles, unless it is to be gathered from the considera.ion that the affinity is so weak that the mixture of fluids preserves its dimensions. What can this weak affinity do, when opposed by a repulsive power of infinite superiority?

In discussing the doctrines of elastic fluids mixed with vapour, Mr. Murray seems disposed to question the accuracy of the fact, that the quantity of vapour is the same in vacuo as in air, though he has not attempted to ascertain in which case it more abounds. This is certainly the touchstone of the mechanical and chemical theories; and I had thought that whoever admitted the truth of the fact, must unavoidably adopt the mechanical theory. Berthollet however, convinced from his own experience, that the fact was incontrovertible, attempts to reconcile it, inimical as it is, to the chemical theory; with what success it is left to others to judge. Mr. Murray joins with Berthollet in condemning as extravagant the position which I maintain, that if the atmosphere were annihilated, we should have little more aqueous vapour than at present exists in it. Upon which I shall only remark, that if either of those gentlemen will calculate, or give a rough estimate upon their hypothesis, of the quantity of aqueous vapour that would be collected around the earth, on the said supposition, I will engage to discuss the subject with them more at large.

In 1802, Dr. Henry announced a very curious and important discovery, which was afterwards published in the Philosophical Transactions; namely, *that the quantity of any gas absorbed by water is increased in direct proportion to the pressure of the gas on the surface of the water.* Previously to this, I was engaged in an investigation of the quantity of carbonic acid in the atmosphere; it was matter of surprise to me that lime water should so readily manifest the presence of carbonic acid in the air, whilst pure water by exposure for any length of time, gave not the least traces of that acid. I thought that length of time ought to compensate for weakness of affinity. In pursuing the subject I found that the quantity of this acid taken up by water was greater or less in proportion to its greater or less density in the gaseous mixture, incumbent upon the surface, and therefore ceased to be surprised at water absorbing so insensible a portion from the atmosphere. I had not however entertained any suspicion that this law was generally applicable to the gases till Dr. Henry's discovery was announced. Immediately upon this, it struck me as essentially necessary in ascertaining the quantity of any gas which a given volume of water will absorb, that we must be careful the gas is perfectly pure or unmixed with any other gas whatever; otherwise the maximum effect for any given pressure cannot be produced. This thought was suggested to Dr. Henry, and found to be correct; in consequence of which, it became expedient to repeat some of his experiments relating to the quantity of gas absorbed under a given pressure. Upon due consideration of all these phenomena, Dr. Henry became convinced, that there was no system of elastic fluids which gave so simple, easy and intelligible a solution of them, as the one I adopt, namely, that each gas in any mixture exercises a distinct pressure, which continues the same if the other gases are withdrawn. In the 8th Vol. of Nicholson's Journal, may be seen a letter addressed to me, in which Dr. Henry has clearly pointed out his reasons for giving my theory a preference.

In the 9th Vol. is a letter from Mr. Gough, containing some animadversions, which were followed by an appropriate reply from Dr. Henry.

In the 8th, 9th, and 10th Volumes of Nicholson's Journal, and in the first Vol. of the Manchester Memoirs *(new series)* may be seen some animadversions of Mr. Gough, on my doctrine of mixed gases, with some of his own opinions on the same subject. Mr. Gough conceives the atmosphere to be a chemical compound of gases, vapour, &c. and he rests his belief chiefly upon the observance of certain hygrometrical phenomena, such as that air absorbs moisture from bodies in certain cases, and in others restores it to them, shewing that air has an affinity for water, which may be overcome by another more powerful one. This opinion, as Mr. Murray observes, is the one we have from Dr. Halley; it was supported by Le Roy, Hamilton and Franklin, and might be considered as the prevailing opinion, till Saussure, in his celebrated Essays on hygrometry, published in 1783, suggested that water was first changed into vapour, and was in that state dissolved by the air. This amphibious theory of Saussure does not seem to have gained any converts to it, though it pointed out the instability of the other. Finally, the theory of the chemical solution of water in air, received its death blow in 1791, by the publication of Pictet's Essay on Fire, and more particularly by De Luc's paper on evaporation, published in the Philosophical Transactions for 1792. These gentlemen demonstrated, that all the train of hygrometrical phenomena takes place just as well, indeed rather quicker, in a vacuum, than in air, provided the same quantity of moisture is present. All the influence that any kind or density of air has, is to retard the effect; but in the end it becomes the same.

The only objection which Mr. Gough has presented that appears to me to raise any difficulty, is that in regard to the propagation of sound: If the atmosphere consist chiefly of two distinct elastic media, it is urged that distant sounds ought to be heard double; that is, the same sound would be heard twice, according as it was brought by one or other of the atmospheres. By calculation I find that if sound move at the rate of 1000 feet per second in an atmosphere of azotic gas, it ought to move in the other gases as follows: namely,

	Feet.	
Sound moves in azotic gas	1000	per second.
——— oxygen gas	930	———
——— carb. acid	804	———
——— aqueous vap.	1175	———

According to this table, if a strong and loud sound were produced 13 miles off, the first would be a weak impression of it brought by the atmosphere of aqueous vapour, in 59 seconds; the second would be the strongest of all, brought by the atmosphere of azotic gass, in 68½ seconds; the third would be much inferior to the second, brought by the oxygenous atmosphere, in 74 seconds; the fourth and last brought by the carbonic acid atmosphere would be extremely weak, in 85 seconds.—Now though observation does not perfectly accord with the theory in this respect, it comes as near it, perhaps, as it does to that of the more simple constitution of the atmosphere which Mr. Gough maintains. Derham, who has perhaps made the greatest number of accurate observations on distant sounds, remarked that the report of a cannon fired at the distance of 13 miles from him, did not strike his ear with a single sound, but that it was repeated 5 or 6 times close to each other. "The two first cracks were louder than the third; but the last cracks were lounder than any of the rest." Cavallo, in his experimental philosophy, after quoting the above observations, proceeds, "this repetition of the sound probably originated from the reflection of a single sound, from hills, houses, or other objects, not much distant from the cannon. But it appears from general observation, and where no echo can be suspected, that the sound of a cannon, at the distance of 10 or 20 miles, is different from the sound when near. In the latter case, the crack is loud and instantaneous, of which we cannot appreciate the height. Whereas in the former case, viz. at a distance, it is a grave sound, which may be compared to a determinate musical sound; and instead of being instantaneous, it begins softly, swells to its greatest loudness, and then dies away growling.—Nearly the same thing may be observed with respect to a clap of thunder, other sounds are likewise altered in quality by the distance." (Vol. 2. page 331.)

I shall now proceed to give my present views on the subject of mixed gases, which are somewhat different from what they were when the theory was announced, in consequence of the fresh lights which succeeding experience has diffused. In prosecuting my enquiries into the nature of elastic fluids, I soon perceived it was necessary, if possible, to ascertain whether the atoms or ultimate particles of the different gases are of the same size or volume in like circumstances of temperature and pressure. By the size or volume of an ultimate particle, I mean in this place, the space it occupies in the state of a pure elastic fluid; in this sense the bulk of the particle signifies the bulk of the supposed impenetrable nucleus, together with that of its surrounding repulsive atmosphere of heat. At the time I formed the theory of mixed gases, I had a confused idea, as many have, I suppose, at this time, that the particles of elastic fluids are all of the same size; that a given volume of oxygenous gas contains just as many particles as the same volume of hydrogenous; or if not, that we had no data from which the question could be solved. But from a train of reasoning, similar to that exhibited at page 71, I became convinced that different gases have *not* their particles of the same size: and that the following may be adopted as a maxim, till some reason appears to the contrary: namely,—

That every species of pure elastic fluid has its particles globular and all of a size; but that no two species agree in the size of their particles, the pressure and temperature being the same.

There was another thing concerning which I was dubious; whether heat was the cause of repulsion. I was rather inclined to ascribe repulsion to a force resembling magnetism, which acts on one kind of matter, and has no effect on another. For, if heat were the cause of repulsion, there seemed no reason why a particle of oxygen should not repel one of hydrogen with the same force as one of its own kind, especially if they were both of a size. Upon more mature consideration, I see no sufficient reason for discarding the common opinion, which ascribes repulsion to heat; and I think the phenomena of mixed gases may still be accounted for, by repulsion, without the postulatum, that their particles are mutually inelastic, and free from such of the preceding objections as I have left unanswered.

When we contemplate upon the disposition of the globular particles in a volume of pure elastic fluid, we perceive it must be analogous to that of a square pile of shot; the particles must be disposed into horizontal strata, each four particles forming a square: in a superior stratum, each particle rests upon four particles below, the points of its contact with all four being 45° above the horizontal plane, or that plane which passes through the centres of the four particles. On this account the pressure is steady and uniform throughout. But when a measure of one gas is presented to a measure of another in any vessel, we have then a surface of elastic globular particles of one size in contact with an equal surface of particles of another: in such case the points of contact of the heterogeneous particles must vary all the way from 40° to 90°; an intestine motion must arise from this inequality, and the particles of one kind be propelled amongst those of the other. The same cause which prevented the two elastic surfaces from maintaining an equilibrium, will always subsist, the particles of one kind being from their size unable to apply properly to the other, so that no equilibrium can ever take place amongst the heterogeneous particles. The intestine motion must therefore continue till the particles arrive at the opposite surface of the vessel against any point of which they can rest with stability, and the equilibrium at length is acquired when each gas is uniformly diffused through the other. In the open atmosphere no equilibrium can take place in such case till the particles have ascended so far as to be restrained by their own weight;

that is, till they constitute a distinct atmosphere.

It is remarkable that when two equal measures of different gases are thus diffused, and sustain an invaried pressure, as that of the atmosphere, the pressure upon each particle after the mixture is less than before. This points out the active principle of diffusion; for, particles of fluids are always disposed to move to that situation where the pressure is least. Thus, in a mixture of equal measures of oxygen and hydrogen, the common pressure on each particle before mixture being denoted by 1, that after the mixture when the gas becomes of half its density, will be denoted by $\sqrt[3]{\frac{1}{2}} = .794$.

This view of the constitution of mixed gases agrees with that which I have given before, in the two following particulars, which I consider as essential to every theory on the subject to give it plausibility.

1st. The diffusion of gases through each other is effected by means of the repulsion belonging to the homogenous particles; or to that principle which is always energetic to produce the dilatation of the gas.

2d. When any two or more mixed gases acquire an equilibrium, the elastic energy of each against the surface of the vessel or of any liquid, is precisely the same as if it were the only gas present occupying the whole space, and all the rest were withdrawn.

In other respects I think the last view accords better with the phenomena, and obviates the objections which Dr. Thomson has brought against the former; particularly in regard to the query, why mixed gases that are known on certain occasions to combine, do not always combine; and why any gaseous particle in its nascent state is more disposed to combination than when it has already assumed the elastic form. It will also more clearly explain the reason of one gas making so powerful and durable a resistance to the entrance of another.

One difficulty still remains respecting vapour, which neither view of the subject altogether removes: though vapour may subsist in the atmosphere upon either supposition, as far as the temperature will admit, not being subject to any more pressure than would arise from its own particles, were the others removed, yet it may be enquired, how does it rise from the surface of water subject to the pressure of the atmosphere? how does vapour which ascends with an elastic force of only half an inch of mercury, detach itself from water when it has the weight of 30 inches of mercury to oppose its ascent? This difficulty applys nearly the same to all theories of the solution of water in air, and it is therefore of consequence for every one, let him adopt what opinion he may, to remove it. Chemical solution but ill explains it; for, the affinity of air for vapour is always described as weak, and yet it is sufficient to overcome the pressure of a powerful force equal to the weight of the atmosphere. I have endeavoured to shew in another place (Manch. Memoirs, Vol. 1. *new series*, page 284) what my own ideas on the subject are. It appears to me, that it is not till the depth of 10 or 12 strata of particles of any liquid, that the pressure upon each perpendicular column becomes uniform; and that several of the particles in the uppermost stratum are in reality subject to but little pressure.

SECTION 3.

ON THE

CONSTITUTION OF LIQUIDS,

And the Mechanical Relations betwixt Liquids and Elastic Fluids.

A liquid or inelastic fluid may be defined to be a body, the parts of which yield to a very small impression, and are easily moved one upon another. This definition may suffice for the consideration of liquids in an hydrostatical sense, but not in a chemical sense. Strictly speaking, there is no substance inelastic; if heat be the cause of elasticity, all bodies containing it must necessarily be elastic: but we commonly apply the word elastic to such fluids only as have the property of condensation in a very conspicuous degree. Water is a liquid or inelastic fluid; but if it is compressed by a great force, it yields a little, and again recovers its original bulk when the pressure is removed. We are indebted to Mr. Canton for a set of experiments by which the compressibility of several liquids is demonstrated. Water, he found, lost about

$\frac{1}{1740}$th part of its bulk by the pressure of the atmosphere.

When we consider the origin of water from steam, we have no reason to wonder at its compressibility, and that in a very small degree; it would be wonderful if water had not this quality. The force of steam at 212° is equal to the pressure of the atmosphere; what a prodigious force must it have when condensed 15 or 18 hundred times? We know that the particles of steam, reduced to the state of water, still retain the greatest part of their heat. What a powerful resistance then ought they not to make against a compressing force? The truth is, water, and by analogy, other liquids, must be considered as bodies, under the controul of two most powerful and energetic agents, attraction and repulsion, between which there is an equilibrium. If any compressing force is applied, it yields, indeed, but in such a manner, as a strong spring would yield, when wound up almost to the highest pitch. When we attempt to separate one portion of liquid from another, the case is different: here the attraction is the antagonist force, and that being balanced by the repulsion of the heat, a moderate force is capable of producing the separation. But even here we perceive the attractive force to prevail, there being a manifest cohesion of the particles. Whence does this arise? It should seem that when two particles of steam coalesce to form water, they take their station so as to effect a perfect equilibrium between the two opposite powers; but if any foreign force intervene, so as to separate the two molecules an evanescent space, the repulsion decreases faster than the attraction, and consequently this last acquires a superiority or excess, which the foreign force has to overcome. If this were not the case, why do they at first, or upon the formation of water, pass from the greater to the less distance?

With regard to the collocation and arrangement of particles in an aggregate of water or any other liquid, I have already observed (page 139) that this is not, in all probability, the same as in air. It seems highly improbable from the phenomena of the expansion of liquids by heat. The law of expansion is unaccountable for, if we confine liquids to one and the same arrangement of their ultimate particles in all temperatures; for, we cannot avoid concluding, if that were the case, the expansion would go on in a progressive way with the heat, like as in air; and there would be no such thing observed as a point of temperature at which the expansion was stationary.

Reciprocal Pressure of Liquids and Elastic Fluids.

When an elastic fluid is confined by a vessel of certain materials, such as wood, earthenware, &c. it is found slowly to communicate with the external air, to give and receive successively, till a complete intermixture takes place. There is no doubt but this is occasioned by those vessels being porous, so as to transmit the fluids. Other vessels, as those of metal, glass, &c. confine air most completely. These therefore cannot be porous; or rather, their pores are too small to admit of the passage of air. I believe no sort of vessel has yet been found to transmit one gas and confine another; such a one is a desideratum in practical chemistry. All the gases appear to be completely porous, as might be expected, and therefore operate very temporarily in confining each other. How are liquids in this respect? Do they resemble glass, or earthenware, or gases, in regard to their power of confining elastic fluids? Do they treat all gases alike, or do they confine some, and transmit others? These are important questions: they are not to be answered in a moment. We must patiently examine the facts.

Before we can proceed, it will be necessary to lay down a rule, if possible, by which to distinguish the *chemical* from the *mechanical* action of a liquid upon an elastic fluid. I think the following cannot well be objected to: *When an elastic fluid is kept in contact with a liquid, if any change is perceived, either in the elasticity or any other property of the elastic fluid, so far the mutual action must be pronounced* CHEMICAL: *but if* NO *change is perceived, either in the elasticity or any other property of the elastic fluid, then the mutual action of the two must be pronounced wholly* MECHANICAL.

If a quantity of lime be kept in water and agitated, upon standing a sufficient time, the lime falls down, and leaves the water transparent: but the water takes a small portion of of the lime which it permanently retains, contrary to the Laws of specific gravity. Why? Because that portion of lime is dissolved by the water. If a quantity of air be put to water and agitated, upon standing a sufficient time, the air rises up to the surface of the water and leaves it transparent; but the water permanently retains a portion of air, contrary to the Laws of specific gravity. Why? Because that small portion of air is dissolved by the water. So far the two explanations are equally satisfactory. But if we place the two portions of water under the receiver of an air pump, and exhaust the incumbent air, the whole portion of air absorbed by the water ascends, and is drawn out of the receiver; whereas the lime remains still in solution as before. If now the question be repeated, why is the air retained in the water? The answer must be, because there is an elastic force on the surface of the water which holds it in. The water appears passive in the business. But, perhaps, the pressure on the surface of the water may have some effect upon its affinity for air, and none on that for lime? Let the air be drawn off from the surfaces of the two portions of water, and another species induced without alleviating the pressure. The lime water remains unchanged; the air escapes from the other much the same as in vacuo. The question of the relation of water to air appears by this fact to be still more difficult; at first the air seemed to be retained by the attraction of the water; in the second case, the water seemed indifferent; in the third, it appears as if r pulsive to the air; yet in all three, it is the same air that has to act on the same water. From these facts, there seems reason then for maintaining three opinions on the subject of the mutual action of air and water; namely, that water attracts air, that water does not attract it, and that water repels air. One of these must be true; but we must not decide hastily. Dr. Priestley once imagined, that the clay of a porous earthen retort, when red hot, "destroys for a time the aerial form of whatever air is exposed to the outside of it; which aerial form it recovers, after it has been transmitted in combination from one part of the clay to another, till it has reached the inside of the retort." But he soon discarded so extravagant an opinion.

From the recent experiments of Dr. Henry, with those of my own, there appears reason to conclude, that a given volume of water absorbs the following parts of its bulk of the several gases.

Bulk of gas absorbed.

1	=	1	Carbonic acid
1	=	1	Sulphuretted hydrogen
1	=	1	Nitrous oxide
$\frac{1}{8}$	=	.125	Olefiant gas
$\frac{1}{27}$	=	.037	Oxygenous gas
$\frac{1}{27}$	=	.037	Nitrous gas
$\frac{1}{27}$	=	.037	Carburetted hydrogen
$\frac{1}{27}$	=	.037	Carbonic oxide ?
$\frac{1}{64}$	=	.0156	Azotic gas
$\frac{1}{64}$	=	.0156	Hydrogenous gas
$\frac{1}{64}$	=	.0156	Carbonic oxide ?

These fractions are the cubes of $\frac{1}{1}$, $\frac{1}{2}$, $\frac{1}{3}$, $\frac{1}{4}$, &c. this shews the distances of the gaseous particles in the water to be always same multiple of the distances without.

In a mixture of two or more gases, the rule holds the same as when the gases are alone; that is, the quantity of each absorbed is the same as if it was the only gas present.

As the quantity of any gas in a given volume is subject to variation from pressure and temperature, it is natural to enquire whether any change is induced in the absorption of these circumstances; the experiments of Dr. Henry have decided this point, by ascertaining, that if the exterior gas is condensed or rarefied in any degree, the gas absorbed is condensed or rarefied in the same degree; so that the proportions absorbed given above are absolute.

One remarkable fact, which has been hinted at is, that no one gas is capable of retaining another in water; it escapes, not indeed instantly, like as in a vacuum, but gradually, like as carbonic acid escapes into the atmosphere from the bottom of a cavity communicating with it.

It remains now to decide whether the relation between water and the abovementioned gases is of a *chemical* or *mechanical* nature. From the facts just stated, it appears evident that the elasticity of carbonic acid and the other two gases of the first class is not at all affected by the water. It remains exactly of the same energy whether the water is present or absent. All the other properties of those gases continue just the same, as far as I know, whether they are alone or blended with water: we must therefore, I conceive, if we abide by the Law just laid down, pronounce the mutual action between these gases and water to be *mechanical.*

A very curious and instructive phenomenon

takes place when a portion of any of the above three gases is thrown up into an eudiometer tube of $\frac{3}{10}$ of an inch diameter over water; the water ascends and absorbs the gas with considerable speed; if a small portion of common air is suddenly thrown up, it ascends to the other, and is commonly separated by a fine film of water for a time. That instant the the two airs come into the above situation, the water suddenly ceases to ascend in the tube, but the film of water runs up with great speed, enlarging the space below, and proportionally diminishing that above, till it finally bursts. This seems to shew that the film is a kind of sieve through which those gases can easily pass, but not common air.

In the other gases it is very remarkable their density within the water should be such as to require the distance of the particles to be just 2, 3 or 4 times what it is without. In olefiant gas, the distance of the particles within is just twice that without, as is inferred from the density being $\frac{1}{8}$. In oxygenous gas, &c. the distance is 3 times as great, and in hydrogenous, &c. 4 times. This is certainly curious, and deserves further investigation; but at present we have only to decide whether the general phenomena denote the relation to be of a chemical or mechanical nature. In no case whatever does it appear that the elasticity of any of these gases is affected; if water takes $\frac{1}{27}$ of its bulk of any gas, the gas so absorbed, exerts $\frac{1}{27}$ of the elasticity, that the exterior gas does, and of course it escapes from the water when the pressure is withdrawn from its surface, or when a foreign one is induced, against which it is not a proper match. As far as is known too, all the other properties of the gases continue the same; thus, if water containing oxygenous gas be admitted to nitrous gas, the union of the two gases is certain; after which the water takes up $\frac{1}{27}$ of its bulk of nitrous gas, as it would have done, if this circumstance had not occurred. It seems clear then that the relation is a *mechanical* one.*

* Dr. Thomson and Mr. Murray have both written largely in defence of the notion that *all* gases are combined with water, that a real union by means of a chemical affinity which water exercises in a greater or less degree towards all gases, takes place; this affinity is supposed to be of the *slight* kind, or of that kind which holds all gases in a state of solution, one amongst another, without any distinction. The oppposite doctrine was first stated in a paper of mine, on the absorption of gases by water. (Manch. Memoirs, *new series*, Vol. 1.) Previously to the publication of that paper, Dr. Henry, who was convinced from his own experience, that the connection of gases with water was of a mechanical nature, wrote two essays in

Carbonic acid gas then presses upon water in the first instance with its whole force; in a short time it partly enters the water, and then the reaction of the part entered, contri-

the 8th and 9th Vol. of Nicholson's Journal, in which the arguments for that opinion are clearly, and, I think, unanswerably stated. I do not intend to enter largely into a discussion of the arguments these gentlemen adopt. Dr. Thomson's leading argument seems to be, that "water will absorb such a portion of each gas, that the repulsion between the particles absorbed, just balances the affinity of water for them." He then proceeds to infer, that the affinity of carbonic acid for water is such as nearly to balance the elasticity, that the affinity of olefiant gas for water is equal to *half* its elasticity, that of oxygen, $\frac{1}{3}$, and of azote $\frac{1}{4}$, &c. Now if a particle of water attract one of carbonic acid by a force analogous to that of repulsion, it must increase directly as the distance decreases; if so, two such particles must be in equilibrium at any distance; and if any other force is applied to the particle of gas propelling it towards the water, the two particles must unite or come into most intimate contact. Hence, I should infer, from Dr. Thomson's principle, that each particle of water would take one of acid, and consequently 1lb. of water would combine with 2½lbs. of carbonic acid nearly. Mr. Murray mentions a great many circumstances which he conceives make against the mechanical hypothesis; for instance, some of the acid and alkaline gases are known to be absorbed largely by water, and undoubtedly by affinity; therefore the less absorbable gases must be under the same influence, only in an inferior degree, and that "it would be impossible to point out the line of distinction between those where the absorption might be conceived to be purely

butes to support the incumbent atmosphere. Finally, the gas gets completely diffused through the water, so as to be of the same density within as without; the gas within the water then presses on the containing vessel only, and reacts upon the incumbent gas. The water then sustains no pressure either from the gas within or without. In olefiant gas the surface of the water supports $\frac{7}{8}$ of the pressure, in oxygenous, &c. $\frac{26}{27}$, and in hydrogenous, &c. $\frac{63}{64}$.

When any gas is confined in a vessel over

mechanical, and those where the exertion of affinity must be allowed to operate." I conceive nothing is more easy than to point out the exact line of distinction: *wherever water is found to diminish or destroy the elasticity of any gas, it is a chemical agent; wherever it does neither of these, it is a mechanical agent.* Whoever undertakes to maintain the chemical theory of the absorption of gases by water, should in the outset overturn the following argument preferred by Dr. Henry: "The quantity of every gas, absorbed by water, follows exactly the ratio of the pressure; and since it is a rule in philosophizing, that effects of the same kind, though differing in degree, are produced by the same cause, it is perfectly safe to conclude, that every, even the minutest portion of any gas, in a state of absorption by water, is retained entirely by incumbent pressure. There is no occasion, therefore, to call in the aid of the law of chemical affinity, when a mechanical law fully and satisfactorily explains the appearances."

water in the pneumatic trough, so as to communicate with the atmosphere through the medium of water, that gas must constantly be filtring through the water into the atmosphere, whilst the atmospheric air is filtring through the water the contrary way, to supply its place in the vessel; so that in due time the air in the vessel becomes atmospheric, as various chemists have experienced. Water in this respect is like an earthenware retort: it admits the gases to go both ways at the same time.

It is not easy to assign a reason why water should be so permeable to carbonic acid, &c. and not to the other gases; and why there should be those differences observable in the others. The densities $\frac{1}{8}$, $\frac{1}{27}$ and $\frac{1}{64}$, have most evidently a reference to a mechanical origin, but none whatever to a chemical one. No mechanical equilibrium could take place if the densities of the gases within were not regulated by this law; but why the gases should not all agree in some one of these forms, I do not see any reason.

Upon the whole it appears that water, like earthenware, is incapable of forming a perfect barrier to any kind of air; but it differs from earthenware in one respect; the last is alike permeable to all the gases, but water is much more permeable to some gases than to others. Other liquids have not been sufficiently examined in this respect.

The mutual action of water, and the greater number of acid gases and alkaline gas partaking most evidently of a chemical nature, will be best considered under the heads of the respective acids and alkalis.

SECTION 4.

ON THE CONSTITUTION OF SOLIDS.

A solid body is one, the particles of which are in a state of equilibrium betwixt two great powers, attraction and repulsion, but in such a manner, that no change can be made in their distances without considerable force. If an approximation of the particles is attempted by force, then the heat resists it; if a separation, then the attraction resists it. The notion of Boscovich of alternating planes of attraction and repulsion seems unnecessary; except that upon forcibly breaking the cohesion of any body, the newly exposed surface must receive such a modification in its atmosphere of heat, as may prevent the future junction of the parts, without great force.

The essential distinction between liquids and solids, perhaps consists in this, that heat changes the figure of arrangement of the ultimate particles of the former continually and gradually, whilst they retain their liquid form; whereas in the latter, it is probable, that change of temperature does no more than change the size, and not the arrangement of the ultimate particles.

Notwithstanding the *hardness* of solid bodies, or the difficulty of moving the particles one amongst another, there are several that admit of such motion without fracture, by the application of proper force, especially if assisted by heat. The ductility and malleability of the metals, need only to be mentioned. It should seem the particles glide along each others surface, somewhat like a piece of polished iron at the end of a magnet, without being at all weakened in their cohesion. The absolute force of cohesion, which constitutes the *strength* of bodies, is an enquiry of great practical importance. It has been found by experiment, that wires of the several metals beneath, being each $\frac{1}{10}$ of an inch in diameter, were just broken by the annexed weights.

Lead............	$29\frac{1}{4}$	Pounds.
Tin	$49\frac{1}{4}$	
Copper.........	$299\frac{1}{4}$	
Brass...........	360	
Silver..........	370	
Iron	450	
Gold	500	

A piece of good oak, an inch square and a yard long, will just bear in the middle 330lbs. But such a piece of wood should not in practice be trusted, for any length of time, with above $\frac{1}{3}$ or $\frac{1}{4}$ of that weight. Iron is about 10 times as strong as oak, of the same dimensions.

One would be apt to suppose that *strength* and *hardness* ought to be found proportionate to each other; but this is not the case. Glass is harder than iron, yet the latter is much the stronger of the two.

Crystallization exhibits to us the effects of the natural arrangement of the ultimate particles of various compound bodies; but we are scarcely yet sufficiently acquainted with chemical synthesis and analysis to understand the rationale of this process. The rhomboidal form may arise from the proper position of 4, 6, 8 or 9 globular particles, the cubic form from 8 particles, the triangular form from 3,

6 or 10 particles, the hexahedral prism from 7 particles, &c. Perhaps, in due time, we may be enabled to ascertain the number and order of elementary particles, constituting any given compound element, and from that determine the figure which it will prefer on crystallization, and *vice versâ*; but it seems premature to form any theory on this subject, till we have discovered from other principles the number and order of the primary elements which combine to form some of the compound elements of most frequent occurrence; the method for which we shall endeavour to point out in the ensuing chapter.

CHAP. III.

ON CHEMICAL SYNTHESIS.

WHEN any body exists in the elastic state, its ultimate particles are separated from each other to a much greater distance than in any other state; each particle occupies the centre of a comparatively large sphere, and supports its dignity by keeping all the rest, which by their gravity, or otherwise are disposed to encroach up it, at a respectful distance. When we attempt to conceive the *number* of particles in an atmosphere, it is somewhat like attempting to conceive the number of stars in the universe; we are confounded with the thought. But if we limit the subject, by taking a given volume of any gas, we seem persuaded that, let the divisions be ever so minute, the number of particles must be finite; just as in a given space of the universe, the number of stars and planets cannot be infinite.

Chemical analysis and synthesis go no farther than to the separation of particles one from another, and to their reunion. No new creation or destruction of matter is within the reach of chemical agency. We might as well attempt to introduce a new planet into the solar system, or to annihilate one already in existence, as to create or destroy a particle of hydrogen. All the changes we can produce, consist in separating particles that are in a state of cohesion or combination, and joining those that were previously at a distance.

In all chemical investigations, it has justly been considered an important object to ascertain the relative *weights* of the simples which constitute a compound. But unfortunately the enquiry has terminated here; whereas from the relative weights in the mass, the relative weights of the ultimate particles or atoms of the bodies might have been inferred, from which their number and weight in various other compounds would appear, in order to assist and to guide future investigations, and to correct their results. Now it is one great object of this work, to shew the importance and advantage of ascertaining *the relative weights of the ultimate particles, both of simple and compound bodies, the number of simple elementary particles which constitute one compound particle, and the number of less compound particles which enter into the formation of one more compound particle.*

If there are two bodies, A and B, which are disposed to combine, the following is the order in which the combinations may take place, beginning with the most simple: namely,

1 atom of A + 1 atom of B = 1 atom of C, binary.
1 atom of A + 2 atoms of B = 1 atom of D, ternary.
2 atoms of A + 1 atom of B = 1 atom of E, ternary.
1 atom of A + 3 atoms of B = 1 atom of F, quaternary.
3 atoms of A + 1 atom of B = 1 atom of G, quaternary.
&c. &c.

The following general rules may be adopted as guides in all our investigations respecting chemical synthesis.

1st. When only one combination of two bodies can be obtained, it must be presumed to be a *binary* one, unless some cause appear to the contrary.

2d. When two combinations are observed, they must be presumed to be a *binary* and a *ternary*.

3d. When three combinations are obtained, we may expect one to be a *binary*, and the other two *ternary*.

4th. When four combinations are observed, we should expect one *binary*, two *ternary*, and one *quaternary*, &c.

5th. A *binary* compound should always be specifically heavier than the mere mixture of its two ingredients.

6th. A *ternary* compound should be specifically heavier than the mixture of a binary and a simple, which would, if combined, constitute it; &c.

7th. The above rules and observations equally apply, when two bodies, such as C and D, D and E, &c. are combined.

From the application of these rules, to the chemical facts already well ascertained, we deduce the following conclusions; 1st. That water is a binary compound of hydrogen and oxygen, and the relative weights of the two elementary atoms are as 1 : 7, nearly; 2d. That ammonia is a binary compound of hydrogen and azote, and the relative weights of the two atoms are as 1 : 5, nearly; 3d. That nitrous gas is a binary compound of azote and oxygen, the atoms of which weigh 5 and 7 respectively; that nitric acid is a binary or ternary compound according as it is derived, and consists of one atom of azote and two of oxygen, together weighing 19; that nitrous oxide is a compound similar to nitric acid, and consists of one atom of oxygen and two of azote, weighing 17; that nitrous acid is a binary compound of nitric acid and nitrous gas, weighing 31; that oxynitric acid is a binary compound of nitric acid and oxygen, weighing 26; 4th. That carbonic oxide is a binary compound, consisting of one atom of charcoal, and one of oxygen, together weighing nearly 12; that carbonic acid is a ternary compound, (but sometimes binary) consisting of one atom of charcoal, and two of oxygen, weighing 19; &c. &c. In all these cases the weights are expressed in atoms of hydrogen, each of which is denoted by unity.

In the sequel, the facts and experiments from which these conclusions are derived, will be detailed; as well as a great variety of others from which are inferred the constitution and weight of the ultimate particles of the principal acids, the alkalis, the earths, the metals, the metallic oxides and sulphurets, the long train of neutral salts, and in short, all the chemical compounds which have hitherto obtained a tolerably good analysis. Several of the conclusions will be supported by original experiments.

From the novelty as well as importance of the ideas suggested in this chapter, it is deemed expedient to give plates, exhibiting the mode of combination in some of the more simple cases. A specimen of these accompanies this first part. The elements or atoms of such bodies as are conceived at present to be simple, are denoted by a small circle, with some distinctive mark; and the combinations consist in the juxta-position of two or more of these; when three or more particles of elastic fluids are combined together in one, it is to be supposed that the particles of the same kind repel each other, and therefore take their stations accordingly.

END OF PART THE FIRST.

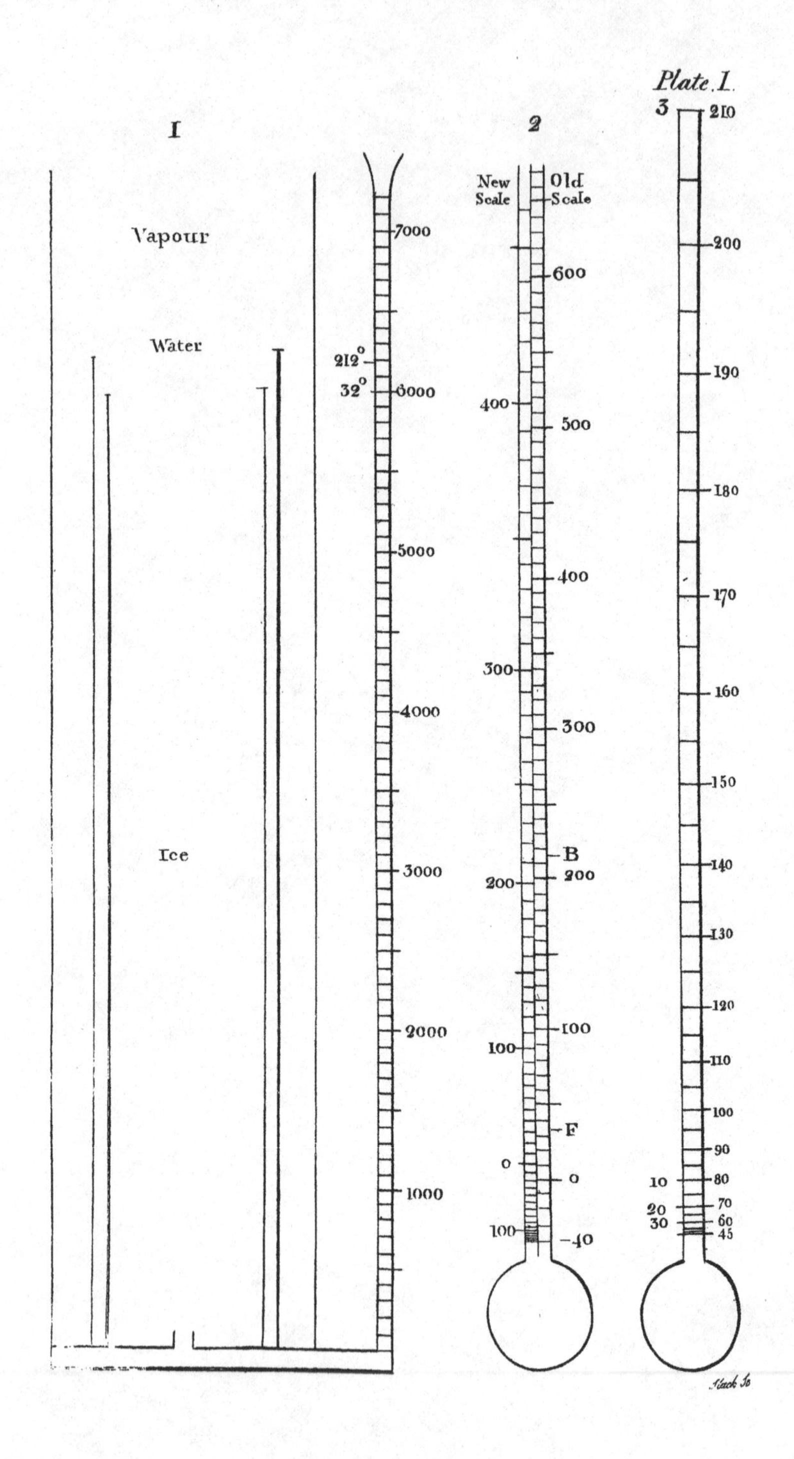
Plate I.
I
Vapour
Water
Ice
212°
32°
7000
6000
5000
4000
3000
2000
1000
2
New Scale
Old Scale
600
500
400
300
B
200
100
F
0
-40
400
300
200
100
0
100
3
210
200
190
180
170
160
150
140
130
120
110
100
90
80
70
60
45
10
20
30

EXPLANATION OF THE PLATES.

PLATE I. Fig. 1. is intended to illustrate the author's ideas on the subject of the capacities of bodies for heat. See page 3. There are three cylindrical vessels placed one within another, having no communication but over their margins; the innermost is connected with a lateral and parallel tube graduated, and supposed to represent the degrees of a thermometer, the scale of which commences at absolute cold; if a liquid (supposed to represent heat) be poured into the tube, it will flow into the inner vessel, through an aperture at the bottom, and rise to the same level in the vessel and the tube. Equal increments of heat in this case are supposed to produce equal increments of temperature. When the temperature has arrived at a certain point (suppose 6000°) the body may be supposed to change its solid form to the liquid, as from ice to water; in which case its capacity for heat is increased, and is to be represented by the second vessel. A considerable portion of liquid must then be poured into the tube before any rise will be perceived, because it flows over the margin of the innermost vessel into the lateral cavity of the second; at length it reaches the level, and then a proportional rise will ensue, till the body becomes converted into an elastic fluid, when the thermometer again becomes stationary—whilst a great portion of heat is entering into the body, now assuming a new capacity.

Fig. 2. is a comparative view of the old and new divisions of the scale of the mercurial thermometer. See Table, page 14. The interval from freezing to boiling water is 180° on both scales, and the extremes are numbered 32° and 212° respectively. There are no other points of temperature in which the two scales can agree.

Fig. 3. is a view of the divisions of a water thermometer, conformably to the new scale of the mercurial; the lowest point is at 45°; the intervals from 45° upwards, to 55°, 65°, 75°, &c. are as the numbers 1, 4, 9, &c. Also, 30° and 60° coincide, as do 20° and 70°, &c.

PLATE II. Fig. 1. represents an air thermometer, or the expansion of air by heat; the numbers are Fahrenheit's, and the intervals are such as represented in the 7th column of the table, at page 14.

Fig. 2. is the logarithmic curve, the ordinates of which are erected at equal intervals, and diminish progressively by the ratio $\frac{4}{5}$. The intervals of the absciss or base of the curve, represent equal intervals of temperature (25° for steam or aqu ous vapour, and 34° for ethereal vapour) the ordinates represent inches of mercury, the weight of which is equal to the force of steam at the temperature. See the 8th and 9th columns of table, at page 14. Thus the force of steam at 212°, and of ethereal vapour at 110°, new scale, is equal to 30 inches of mercury; at 187° the force of steam is half as much, or 15 inches, and at 76°, that of ethereal vapour is also 15 inches, &c.

Fig. 3. is a device suggested by Mr. Ewart, to illustrate the idea which I have developed in the section on the temperature of the atmosphere. It is a cylindrical vessel close at one end and open at the other, having a moveable piston sliding within it: the vessel is supposed to contain air, and a weight is connected with the piston as a counterpoise to it. There is also a thermometer supposed to pass through the side of the vessel, and to be cemented into it. Now if we may suppose the piston to move without friction, and the vessel to be taken up into the atmosphere, the piston will gradually ascend, and suffer the air within to dilate, so as to correspond every where with the exterior air in density. This dilatation tends to diminish the temperature of the air within (provided no heat is acquired from the vessel.) Such an instrument would shew what the theory requires, namely, that the temperature of the air within would every where in the same vertical column agree with that without, though the former would not receive or part with any heat absolutely, or in any manner communicate with the external air.

PLATE III. See page 135.—The balls in Fig. 1 and 2 represent particles of water: in the former, the square form denotes the arrangement in water, the rhomboidal form in the latter, denotes the arrangement in ice. The angle is always 60° or 120°.

Fig. 3. represents the perpendicular section of a ball resting upon two others, as 4 and 8, Fig 1.

Fig. 4. represents the perpendicular section of a ball resting upon two balls, as 7 and 5, Fig. 2. The perpendiculars of the triangles shew the heights of the strata in the two arrangements.

Fig. 5 represents one of the small spiculæ of ice formed upon the sudden congelation of water cooled below the fre zing point. See page 134.

Fig. 6. represents the shoots or ramifications of ice at the commencement of congelation. The angles are 60° and 120°.

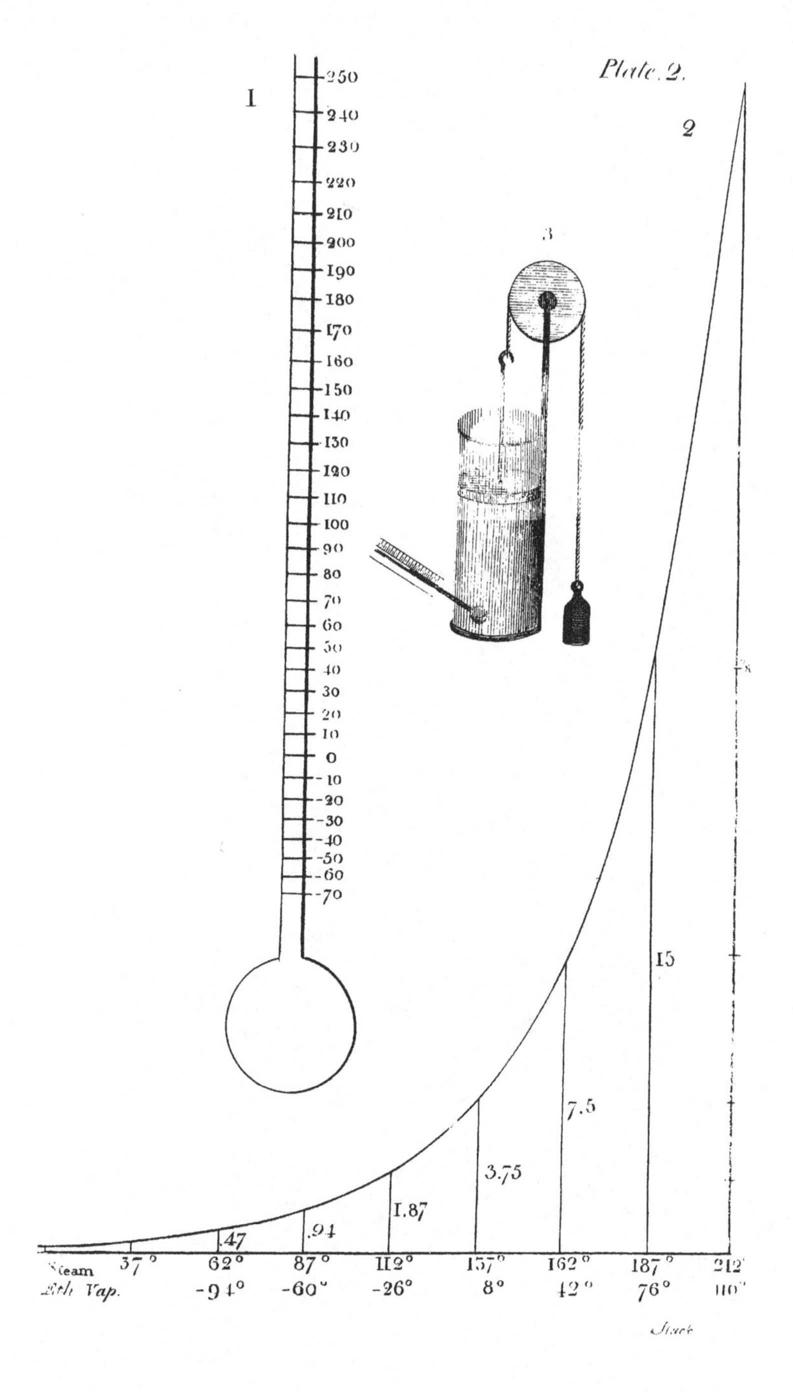
Plate. 2.
I
250
240
230
220
210
200
190
180
170
160
150
140
130
120
110
100
90
80
70
60
50
40
30
20
10
0
-10
-20
-30
-40
-50
-60
-70
2
3
15
7.5
3.75
1.87
.94
.47
Steam 37° 62° 87° 112° 137° 162° 187° 212°
Æth. Vap. -94° -60° -26° 8° 42° 76° 110°

Plate 3.

1

2

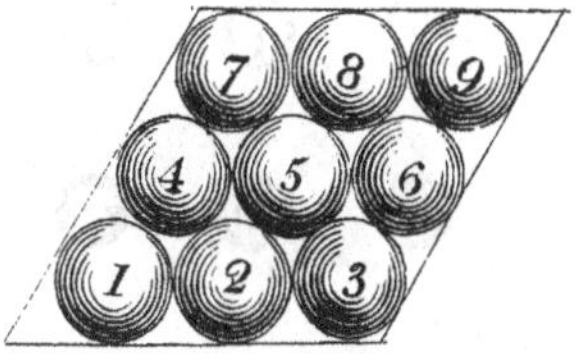

3

4

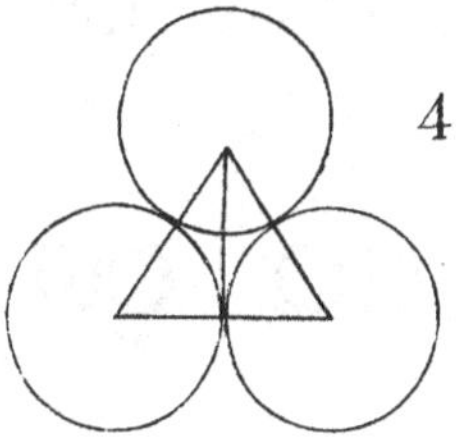

5

6

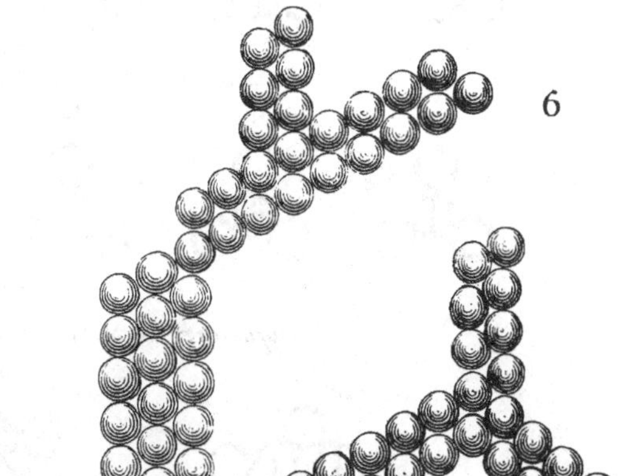

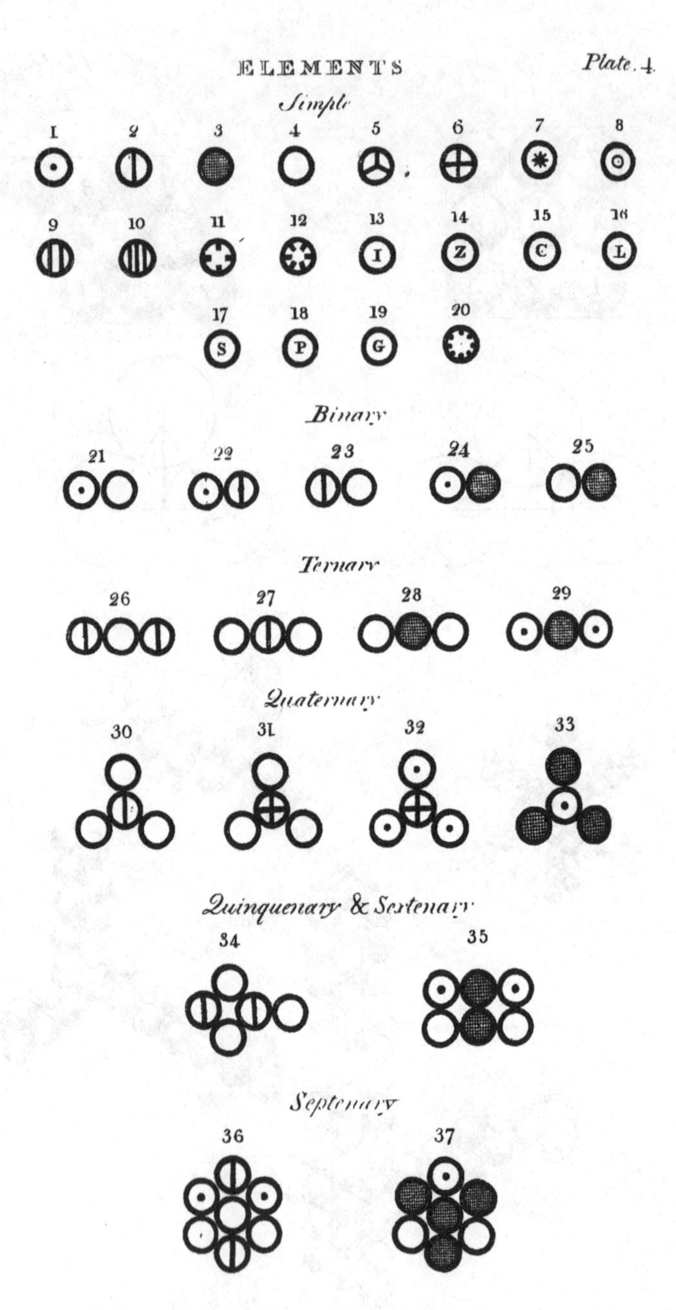
ELEMENTS
Plate. 4.
Simple
1
2
3
4
5
6
7
8
9
10
11
12
13
14
15
16
17
18
19
20
I
Z
C
L
S
P
G
Binary
21
22
23
24
25
Ternary
26
27
28
29
Quaternary
30
31
32
33
Quinquenary & Sextenary
34
35
Septenary
36
37

PLATE IV. This plate contains the arbitrary marks or signs chosen to represent the several chemical elements or ultimate particles.

Fig.			Fig.		
1	Hydrog. its rel. weight	1	11	Strontites - - -	46
2	Azote, - - - -	5	12	Barytes - - - - -	68
3	Carbone or charcoal, -	5	13	Iron - - - - - -	38
4	Oxygen, - - - - -	7	14	Zinc - - - - - -	56
5	Phosphorus, - - -	9	15	Copper - - - - -	56
6	Sulphur, - - - - -	13	16	Lead - - - - - -	95
7	Magnesia, - - - -	20	17	Silver - - - - - -	100
8	Lime, - - - - - -	23	18	Platina - - - - -	100
9	Soda, - - - -	28	19	Gold - - - - - -	140
10	Potash, - - - - -	42	20	Mercury - - - - -	167

21. An atom of water or steam, composed of 1 of oxygen and 1 of hydrogen, retained in physical contact by a strong affinity, and supposed to be surrounded by a common atmosphere of heat; its relative weight = - - - - - - 8

22. An atom of ammonia, composed of 1 of azote and 1 of hydrogen - - - - - - - - - - - 6

23. An atom of nitrous gas, composed of 1 of azote and 1 of oxygen - - - - - - - - - - 12

24. An atom of olefiant gas, composed of 1 of carbone and 1 of hydrogen - - - - - - - - - - 6

25 An atom of carbonic oxide composed of 1 of carbone and 1 of oxygen - - - - - - - - - 12

26. An atom of nitrous oxide, 2 azote + 1 oxygen - 17

27. An atom of nitric acid, 1 azote + 2 oxygen - - 19

28. An atom of carbonic acid, 1 carbone + 2 oxygen 19

29. An atom of carburetted hydrogen, 1 carbone + 2 hydrogen - - - - - - - - - - - - - 7

30. An atom of oxynitric acid, 1 azote + 3 oxygen 26

31. An atom of sulphuric acid, 1 sulphur + 3 oxygen 34

32. An atom of sulphuretted hydrogen, 1 sulphur + 3 hydrogen - - - - - - - - - - - - - 16

33. An atom of alcohol, 3 carbone + 1 hydrogen - 16

34. An atom of nitrous acid, 1 nitric acid + 1 nitrous gas - - - - - - - - - - - - - - - - 31

35. An atom of acetous acid, 2 carbone + 2 water - 26

36. An atom of nitrate of ammonia, 1 nitric acid + 1 ammonia + 1 water - - - - - - - - - 33

37. An atom of sugar, 1 alcohol + 1 carbonic acid - 35

Enough has been given to shew the method; it will be quite unnecessary to devise characters and combinations of them to exhibit to view in this way all the subjects that come under investigation; nor is it necessary to insist upon the accuracy of all these compounds, both in number and weight; the principle will be entered into more particularly hereafter, as far as respects the individual results. It is not to be understood that all those articles marked as simple substances, are necessarily such by the theory; they are only necessarily of such weights. Soda and Potash, such as they are found in combination with acids, are 28 and 42 respectively in weight; but according to Mr. Davy's very important discoveries, they are metallic oxides; the former then must be considered as composed of an atom of metal, 21, and one of oxygen, 7; and the latter, of an atom of metal, 35, and one of oxygen, 7. Or, soda contains 75 per cent. metal and 25 oxygen; potash, 83.3 metal and 16.7 oxygen. It is particularly remarkable, that according to the above-mentioned gentleman's essay on the Decomposition and Composition of the fixed alkalies, in the Philosophical Transactions (a copy of which essay he has just favoured me with) it appears that "the largest quantity of oxygen indicated by these experiments was, for potash 17, and for soda, 26 parts in 100, and the smallest 13 and 19."

DIRECTIONS TO THE BINDER.

Plate 1 to face page 217.
2 to face page 218.
3 to follow plate 2.
4 to face page 219.

PRINTED BY S. RUSSELL, MANCHESTER.

A

NEW SYSTEM

OF

CHEMICAL PHILOSOPHY.

PART II.

BY

JOHN DALTON.

Manchester:

Printed by Russell & Allen, Deansgate,

FOR

R. BICKERSTAFF, STRAND, LONDON.

1810.

TO

HUMPHRY DAVY, Esq. Sec. R. S.

PROFESSOR OF CHEMISTRY IN THE ROYAL INSTITUTION, &c. &c.

AND TO

WILLIAM HENRY, M. D. F. R. S.

VICE PRESIDENT OF THE LITERARY AND PHILOSOPHICAL SOCIETY, MANCHESTER, &c. &c.

THE SECOND PART

OF

THIS WORK IS RESPECTFULLY INSCRIBED,

AS A TESTIMONY TO THEIR DISTINGUISHED MERIT IN THE PROMOTION OF CHEMICAL SCIENCE,

AND

AS AN ACKNOWLEDGMENT OF

THEIR FRIENDLY COMMUNICATIONS AND ASSISTANCE,

BY THE

AUTHOR.

PREFACE.

WHEN the first part of this work was published, I expected to complete it in little more than a year; now two years and a half have elapsed, and it is yet in a state of imperfection. The reason of it is, the great range of experiments which I have found necessary to take. Having been in my progress so often misled, by taking for granted the results of others, I have determined to write as little as possible but what I can attest by my own experience. On this account, the following work will be found to contain more original facts and experiments, than any other of its size, on the elementary principles of chemistry. I do not mean to say that I have copied the minutes of my note-book; this would be almost as reprehensible as writing without any experience; those who are conversant in practical chemistry, know that not more than one new experiment in five is fit to be reported to the public; the rest are found, upon due reflection, to be some way or other defective, and are useful only as they shew the sources of error, and the means of avoiding it.

Finding that my design could not be completed, without a second volume, I was desirous to finish the 5th chapter, which treats of the compounds of two elements, in the part now edited; but the work is enlarged so much, and the time is so far advanced, that I have been obliged to omit two or three important sections, particularly the metallic oxides and sulphurets, which I am aware will demand no inconsiderable share of attention. After these are disposed of, the 6th chapter will treat of compounds of 3 or more elements; this will comprehend the vegetable and other acids not yet noticed, the hydrosulphurets, the neutral salts, compound combustibles, &c. &c.

Whatever may be the result of my plan to render the work somewhat like complete, by the addition of another volume, I feel great present satisfaction in having been enabled thus far to develope that theory of chemical synthesis, which, the longer I contemplate, the more I am convinced of its truth. Enough is already done to enable any one to form a judgment of it. The facts and observations yet in reserve, are only of the same kind as those already advanced; if the latter are not sufficient to convince, the addition of the former will be but of little avail. In the mean time, those who, with me, adopt the system, will, I have no doubt, find it a very useful guide in the prosecution of all chemical investigations.

In the arrangement of the articles treated of, I have endeavoured to preserve order; namely, to take such bodies as are simple, according to our present knowledge; and next, those bodies that are compounds of two elements; but in this I have not always succeeded. For, in some instances, it has not been quite clear what was simple, and what compound; in others, the compounds of three or more elements have been so intimately connected with those of two, that it was found impracticable to give a satisfactory account of the latter, without entering more or less into a description of the former.

In regard to nomenclature, I have generally adopted what was most current; perhaps, in a few instances, my peculiar views may have led me to deviate from this rule. I have called those salts *carbonates*, which are constituted of one atom of carbonic acid united to one of base; and the like for other salts. But some moderns call the *neutral* salts carbonates, and the former *subcarbonates;* whereas, I should call the neutral carbonates of soda and potash *supercarbonates*, consisting of two atoms of acid and one of base. I have, however, continued to call the common *nitrates* by that name, though most of them must be considered on my system as *supernitrates*. I am not very anxious upon this head, as it is evident that if the system I proceed upon be adopted, a general reformation of nomenclature will be the consequence, having a reference to the *number of atoms*, as well as to the *kind of elements*, constituting the different compound bodies.

Nov. 1810.

CONTENTS OF PART SECOND.

NEW SYSTEM

OF

CHEMICAL PHILOSOPHY.

PART II.

CHAP. IV.

ON ELEMENTARY PRINCIPLES.

IN order to convey a knowledge of chemical facts and experience the more clearly, it has been generally deemed best to begin with the description of such principles or bodies as are the most simple, then to proceed to those that are compounded of two simple elements, and afterwards to those compounded of three or more simple elements. This plan will be kept in view in the following work, as far as is convenient. By elementary principles, or simple bodies, we mean such as have not been decomposed, but are found to enter into combination with other bodies. We do not know that any one of the bodies denominated elementary, is absolutely indecomposable; but it ought to be called simple, till it can be analyzed. The principal simple bodies are distinguished by the names *oxygen*, *hydrogen*, *azote* or *nitrogen*, *carbone* or *charcoal*, *sulphur*, *phosphorus*, and the *metals*. The fixed alkalis and the earths were lately undecomposed; but it has long been suspected that they were compounds; and Mr. Davy has recently shewn, by means of galvanic agency, that some of them contain metals, and have all the characters of metallic oxides; no harm can arise, it is conceived, therefore, from placing all the earths in the same class as the metallic oxides.

After the elementary or simple bodies, those compounded of two elements require next to be considered. These compounds form a highly interesting class, in which the new principles adopted are capable of being exhibited, and their accuracy investigated by direct experiment. In this class we find several of the most important agents in chemistry; namely, water, the sulphuric, nitric, muriatic, carbonic and phosphoric acids, most of the compound gases, the alkalis, earths, and metallic oxides.

In the succeeding classes we shall find the

more complex compounds to consist of 3, 4, or more elementary principles, particularly the salts; but in these cases, it generally happens that one compound atom unites to one simple atom, or one compound to another compound, or perhaps to two compound atoms; rather than 4 or 6 simple elementary atoms uniting in the same instant. Thus the law of chemical synthesis is observed to be simple, and always limited to small numbers of the more simple principles forming the more compound.

SECTION 1.

OXYGEN.

The most simple state in which oxygen can be procured, is that of a gas or elastic fluid. The gas may be obtained,

1st. *Without the application of heat.* Put 2 ounces of red lead (minium) into a 5 ounce gas bottle; to which put one ounce of the strongest sulphuric acid; then instantly shake it a little to promote mixture, and apply the stopper with a bent tube: suddenly a great heat is generated, white fumes fill the bottle, and a copious flow of gas ensues, which may be received in phials over water, in the usual way. About 30 cubic inches of gas may be expected. This gas should be exposed to a mixture of lime and water, which absorbs about $\frac{1}{5}$ of it (carbonic acid), and leaves the rest nearly pure.

2. *With the application of heat.* Put 2 ounces of manganese (the common black oxide) into an iron bottle, or gun barrel properly prepared, to which a recurved tube is adapted. This is then to be put into a fire, and heated red; oxygenous gas will come over, and may be received as before; it usually contains a small portion of carbonic acid, which may be extracted by lime water. Three or four pints of air may thus be obtained.

3. Two ounces of manganese may be put into a phial, with the same weight of sulphuric acid; the mixture being made into a paste, apply the heat of a candle or lamp, and the gas comes over as before, nearly pure, if taken over water.

4. If an ounce of nitre be put into an iron bottle, and exposed to a strong red heat, a large quantity of gas (2 or 3 gallons) may be obtained. It consists of about 3 parts oxygen and 1 azote, mixed together.

5. Put 100 grains of the salt called oxymuriate of potash into a glass or earthenware

retort; apply the heat of a lamp, &c. till the retort grows nearly red, and a quantity of oxygenous gas will come over with great rapidity. About 100 cubic inches will be obtained, free from carbonic acid, and in other respects very pure.

Various other methods are occasionally used to obtain this gas, but the above are the principal; and for one who has not had much experience, or who wants only a small quantity of gas nearly pure, the first and second are the easiest and most economical.

Properties of Oxygen.

To enumerate all the properties of oxygen, and the combinations into which it enters, would be to write one half of a treatise on chemistry. It will be sufficient, under the present head, to point out some of its more distinguishing features.

1. If the specific gravity of atmospheric air be denoted by 1, that of oxygen will be 1.127 according to Davy, but some have found it rather less. One hundred cubic inches of it, at the temperature 55°, and pressure 30 inches of mercury, weigh nearly 35 grains; the same quantity of atmospheric air weighs 31.1 grains. The weight of an atom of oxygen is denoted

by 7, that of an atom of hydrogen being 1; this is inferred from the relative weights of those elements entering into combination to form water. The diameter of a particle of oxygen, in its elastic state, is to that of one of hydrogen, as .794 to 1.*

2. Oxygen unites with hydrogen, charcoal, azote, phosphorus, and other bodies denominated combustible, and that in various manners and proportions; when mixed with hydrogen and some other elastic fluids, it explodes by an electric spark, with noise, and a violent concussion of the vessel, together with the extrication of much heat. This is called *detonation*. In other cases, the union of oxygen with bodies is more slow, but accompanied by heat. This is usually called *combustion*, as in the burning of charcoal; and *inflammation*, when accompanied with flame, as in the burning of *oil*.—In other cases, the union is still more slow, and consequently with

* For, the diameter of an elastic particle is as $\sqrt[3]{}$ (weight of one atom ÷ specific gravity of the fluid). Whence, denoting the weight of an atom of hydrogen by 1, and the specific gravity of hydrogenous gas also by 1, the weight of an atom of oxygen will be 7, and the specific gravity of oxygenous gas, 14; we have then $\sqrt[3]{\frac{7}{14}}$: 1, or $\sqrt[3]{\frac{1}{2}}$: 1, or .794 : 1 : : diameter of an atom of oxygen : the diameter of one of hydrogen.

little increase of temperature, as in the *rusting* of metals. This is called *oxidation*.

Bodies burn in the atmosphere, or air surrounding the earth, in consequence of the oxygen it contains, which is found to be rather more than $\frac{1}{5}$th of the whole mass. Hence it is not surprising, that in pure oxygen they burn with a rapidity and splendor far superior to what is observed in ordinary combustion. This is easily exhibited, by plunging the ignited body into a large phial full of oxygen; a taper, small iron wire, charcoal, and above all phosphorus, burns with inconceivable brilliancy in this gas.—The nature of the new compounds formed, will be best considered after the properties of the other elementary principles have been enumerated.

3. That part of the atmosphere which is necessary to the support of animal life, is oxygenous gas. Hence, an animal can subsist much longer in a given quantity of pure oxygenous gas, than in the same quantity of common or atmospheric air. In the process of respiration, a portion of oxygenous gas disappears, and an equal one of carbonic acid is produced; a similar change takes place in the combustion of charcoal; hence it is inferred, that respiration is the source of animal heat. Atmospheric air inspired, contains about 21

per cent. oxygenous gas; the air expired, usually contains about 17 per cent. oxygen, and 4 carbonic acid. But if a full expiration of air be made, and the last portion of the expired air be examined, it will be found to have 8 or 9 per cent. carbonic acid, and to have lost the same quantity of oxygenous gas.

4. Oxygenous gas is not sensibly affected by continually passing electric sparks or shocks through it; nor has any other operation been found to decompose it.

SECTION 2.

HYDROGEN.

Hydrogenous gas may be procured by taking half an ounce of iron or zinc filings, turnings, or other small pieces of these metals, putting them into a phial, with two or three ounces of water, to which pour one quarter as much sulphuric acid, and an effervescence will be produced, with abundance of the gas, which may be received over water in the usual way.

Some of its distinguishing properties are:

1. It is the lightest gas with which we are acquainted. Its specific gravity is nearly .0805, that of atmospheric air being 1. This

is nearly the mean attained from the results of different philosophers. Whence we find, that 100 cubic inches of this gas weigh nearly $2\frac{1}{2}$ grains at the mean temperature and pressure. It may be stated to be $\frac{1}{14}$th of the weight of oxygen, and $\frac{1}{12}$th that of azote, and nearly the same fractional part of the weight of common air. The weight of an atom of hydrogen is denoted by 1, and is taken for a standard of comparison for the other elementary atoms. The diameter of an atom of hydrogen, in its elastic state, is likewise denoted by unity, and considered as a standard of comparison for the diameters of the atoms of other elastic fluids.

2. It extinguishes burning bodies, and is fatal to animals that breathe it.

3. If a phial be filled with this gas, and a lighted taper, or red hot iron, be brought to its mouth, the gas will take fire, and burn gradually till the whole is consumed. The flame is usually reddish, or yellowish white.

4. When oxygen and hydrogen gas are mixed together, no change is perceived; but if a lighted taper is brought to the mixture, or an electric spark passed through it, a violent explosion takes place. The two gases unite in a proportion constantly the same, and produce steam, which in a cold medium is in-

stantly condensed into water. When 2 measures of hydrogen are mixed with 1 of oxygen, and exploded over water, the whole gas disappears, and the vessel becomes filled with water, in consequence of the formation and subsequent condensation of the steam.

If 2 measures of atmospheric air be mixed with 1 of hydrogen, and the electric spark made to pass through the mixture, an explosion ensues, and the residuary gas is found to be 1$\frac{3}{4}$ measures, consisting of azote and a small portion of hydrogen. The portion of the mixture which disappears, 1$\frac{1}{4}$, being divided by 3, gives 42 nearly, denoting the oxygen in two measures of atmospheric air, or 21 per cent. The instrument for exploding such mixtures in is called *Volta's eudiometer*.

5. Another remarkable property of hydrogen deserves notice, though it is not peculiar to it, but belongs in degree to all other gases that differ materially from atmospheric air in specific gravity; if a cylindrical jar of 2 or more inches in diameter, be filled with hydrogen, placed upright and uncovered for a moment or two, nearly the whole will vanish, and its place be supplied by atmospheric air. In this case it must evidently leave the vessel in a body, and the other enter in the same manner. But if the jar of hydrogen be held with its

mouth downwards, it slowly and gradually wastes away, and atmospheric air enters in the same manner; after several minutes there will be found traces of hydrogen remaining in the jar. If a tube of 12 inches long and $\frac{1}{4}$ inch internal diameter, be filled with hydrogen, there is little difference perceived whether it is held up or down; the gas slowly and gradually departs in each case, and as much may be found after 10 minutes have expired, as would be after 2 or 3 seconds if the tube were an inch or more in diameter. If a 3 or 4 ounce phial be filled with hydrogen, and a cork adapted, containing a tube of 2 or 3 inches long and $\frac{1}{10}$ inch internal diameter, it does not make any material difference in the waste of the gas whether the phial is held up or down; it will be some hours before the hydrogen gets dispersed.

6. Hydrogen gas bears electrification without any change.

SECTION 3.

ON AZOTE OR NITROGENE.

Azotic or nitrogene gas may be procured from atmospheric air, of which it constitutes the greater part, by various processes: 1st. To

100 measures of atmospheric air put 80 of nitrous gas; the mixture having stood some time, must be passed two or three times through water; it will still contain a small portion of oxygen; to the residuum put 5 more measures of nitrous gas, and proceed as before; small portions of the residuum must then be tried separately, by nitrous gas and by atmospheric air, to see whether any diminution takes place; whichever produces a diminution after the mixture, shews that it is wanting, and the other redundant; consequently a small addition to the stock must be made accordingly. By a few trials the due proportion may be found, and the gas being then well washed, may be considered as pure azotic. 2. If a quantity of liquid sulphuret of lime (a yellow liquid procured by boiling one ounce of a mixture of equal parts sulphur and lime in a quart of water, till it becomes a pint) be agitated in 2 or 3 times its bulk of atmospheric air for some time, it will take out all the oxygen, and leave the azotic gas pure. 3. If to 100 measures of atmospheric air, 42 of hydrogen be put, and an electric spark passed through the mixture, an explosion will take place, and there will be left 80 measures of azotic gas, &c.

The properties of this gas are;—

1. The specific gravity of azotic gas at the

temperature of 55° and pressure 30 inches, is .967 according to Davy, that of air being 1. The weight of 100 cubic inches is nearly 30 grains. The weight of an atom of azote is denoted by 5, that of an atom of hydrogen being 1; this is inferred chiefly from the compound denominated ammonia, and from those of azote and oxygen, as will be seen hereafter. The diameter of a particle of azote in its elastic state, is to that of one of hydrogen, as .747 to 1.

2. Like hydrogen, it extinguishes burning bodies, and is fatal to animals that breathe it.

3. Azotic gas is less prone to combination than most, if not all, other gases; it never combines with any other gas simply of itself; but if a mixture of it and oxygen has the electric spark passed through it for a long continuance, a slow combustion of the azote takes place, and nitric acid is formed. In other cases azote may be obtained in combination with oxygen in various proportions, and the compounds can be analyzed, but are not so easily formed in the synthetic way.

4. Azotic gas, as has been noticed, constitutes nearly $\frac{4}{5}$ths of atmospheric air, notwithstanding its being fatal to animals that breathe it in its unmixed state; the other $\frac{1}{5}$th is oxygenous gas, which is merely mixed with and

diffused through the former, and this mixture constitutes the principal part of the atmosphere, and is suited, as we perceive, both for animal life and combustion.

5. Azotic gas is not affected by repeated electrification.

SECTION 4.

ON CARBONE OR CHARCOAL.

If a piece of wood be put into a crucible, and covered with sand, and the whole gradually raised to a red heat, the wood is decomposed; water, an acid, and several elastic fluids are disengaged, particularly carbonic acid, carburetted hydrogen, and carbonic oxide. Finally, there remains a black, brittle, porous substance in the crucible, called *charcoal*, which is incapable of change by heat in close vessels, but burns in the open air, and is converted into an elastic fluid, carbonic acid. Charcoal constitutes from 15 to 20 per cent. of the weight of the wood from which it was derived.

Charcoal is insoluble in water; it is without taste or smell, but contributes much to correct putrefaction in animal substances. It is less liable to decay than wood by the action of air and water. When new, it gradually absorbs moisture from the atmosphere, amounting to 12 or 15 per cent. of its weight. One half of the moisture may be expelled again by the heat of boiling water, if long continued ; the other requires a higher temperature, and then carries with it a portion of charcoal. I took 350 grains of charcoal that had been exposed to the atmosphere for a long time ; this was subjected to the heat of boiling water for one hour and a half ; it lost 7 grains in the first quarter of an hour, 6 in the second, and finally it had lost 25 grains.

Several authors have maintained that charcoal, after being heated red, has the property of absorbing most species of elastic fluids, in such quantities as to exceed its bulk several times ; by which we are to understand a chemical union of the elastic fluids with the charcoal. The results of their experiments on this head, are so vague and contradictory, as to leave little credit even to the fact of any such absorption. I made 1500 grains of charcoal red hot, then pulverized it, and put it into a Florence flask with a stopcock ; to this a bladder filled with carbonic acid was connected ; this experiment was continued for a week, and occasionally examined by weighing the flask and its contents. At first there appeared an increase of weight of 6 or 7 grains, from the acid mingling with the common air in the flask, of less specific gravity; but the succeeding increase was not more than 6 grains, and arose from the moisture which permeated the bladder: for the bladder continued as distended as at first, and finally upon examination was found to contain nothing but atmospheric air. Yet carbonic acid is stated to be the most absorbable by charcoal. One of the authors above alluded to, asserts that the heat of boiling water is sufficient to expel the greater part of the gases so absorbed. Now this is certainly not true, as Allen and Pepys have shewn; and most practical chemists know that no air is to be obtained from moist charcoal below a red heat. Hence the weight acquired by fresh made charcoal, is in all probability to be wholly ascribed to the moisture which it absorbs from the atmosphere; and it is to the decomposition of this water, and the union of its elements with charcoal, that we obtain such an abundance of gases by the application of a red heat.

It was the prevailing opinion some time ago that charcoal was an oxide of diamond; but Mr. Tennant, and more recently Messrs. Allen and Pepys, have shewn that the same quantity of carbonic acid is obtained from the combustion of the diamond as from that of an equal weight of charcoal; we must therefore conclude, that the diamond and charcoal are the same element in different states of aggregation.

Berthollet contends that charcoal contains hydrogen; this doctrine is farther countenanced by some experiments of Berthollet jun. in the Annales de Chimie, Feb. 1807; Mr. Davy's experience seems also on the same side. But their observations do not appear to me to warrant any other conclusion than that it is extremely difficult to obtain and operate upon charcoal entirely free from water. Hydrogen appears no more essential to charcoal than air is essential to water.

From the various combinations of charcoal with other elements hereafter to be mentioned, the weight of its ultimate particle is deduced to be 5, or perhaps 5.4, that of hydrogen being denoted by unity.

Charcoal requires a red heat, just visible by day light, to burn it: this corresponds to 1000° of Fahrenheit nearly.

SECTION 5.

ON SULPHUR.

Sulphur or brimstone is an article well known; it is an element pretty generally disseminated, but is most abundant in volcanic countries, and in certain minerals. A great part of what is used in this country is imported from Italy and Sicily; the rest is obtained from the ores of copper, lead, iron, &c.

Sulphur is fused by a heat a little above that of boiling water. It is usually run into cylindrical molds, and upon cooling becomes *roll* sulphur. In this case the rolls become highly electrical by friction: they are remarkably brittle, frequently falling in pieces by the contact of the warm hand. Its specific gravity is 1.98 or 1.99.

Sulphur is sublimed by a heat more than sufficient to fuse it; the sublimate constitutes the common *flowers* of sulphur. The effects of the different gradations of heat on sulphur are somewhat remarkable. It is fused at 226° or 228° of Fahrenheit, into a thin fluid; it begins to grow thick, darker, and viscid at about 350°, and continues so till 600° or upwards, the fumes becoming gradually more copious. This viscid mass, if poured into water, continues to retain a degree of tenacity after being cooled ; but finally it becomes of a hard and smooth texture, much less brittle than common roll sulphur.

For any thing certainly known yet, sulphur appears to be an elementary substance. It enters into composition with various bodies ; and from a comparison of several compounds, I deduce the weight of an atom of sulphur to be nearly 14 times that of hydrogen ; it is possible it may be somewhat more or less, but I think the error of the above cannot exceed 2. Mr. Davy seems to conclude, from galvanic experiments on sulphur, that it contains oxygen ; this may be the case, from the great weight of the elementary particles ; but it should contain 50 per cent. oxygen, or none at all.

Berthollet jun. seems to conclude that sulphur contains hydrogen (Annal. de Chimie, Feb. 1807). Mr. Davy inclines to this idea (Philos. Transac. 1807). That some traces of hydrogen may be discovered in sulphur there cannot be much doubt. Dr. Thomson has well observed the difficulty of obtaining sulphur free from sulphuric acid ; but if sulphuric acid be present, water must also be found, and consequently hydrogen. A strong argument against the existence of hydrogen as an essential in sulphur, is derived from the consideration of the low specific heat of sulphur. If this article contained 7 or 8 per cent. of hydrogen, or 50 per cent. of oxygen, or as much water, it would not have the low specific heat of .19.

Sulphur burns in the open air at the temperature of 500°; it unites with oxygen, hydrogen, the alkalis, earths and metals, forming a great variety of interesting compounds, which will be considered in their respective places.

SECTION 6.

ON PHOSPHORUS.

Phosphorus is an article having much the same appearance and consistency as white wax. It is usually prepared from the bones of animals, which contain one of its compounds, phosphate of lime, by a laborious and complex process. The bones are calcined in an open fire; when reduced to powder, sulphuric acid diluted with water is added; this acid takes part of the lime, and forms an insoluble compound, but detaches superphosphate of lime, which is soluble in water. This solution is evaporated, and the salt is obtained in a glacial state. The solid is reduced to powder, and mixed with half its weight of charcoal; then the mixture is put into an earthenware retort, and distilled by a strong red heat, when the phosphorus comes over, and is received in the water into which the tube of the retort is immersed.

Phosphorus is so extremely inflammable, that it is required to be preserved in water: It melts about blood heat; and in close vessels it can be heated up to 550°, when it boils, and of course distils. When exposed to the air, it undergoes slow combustion; but if heated to 100° or upwards, it is inflamed, burns with rapidity and the emission of great heat, accompanied with white fumes. It combines with oxygen, hydrogen, sulphur and other combustible bodies, and with several of the metals.

Phosphorus is soluble in expressed and other oils, in alcohol, ether, &c.; these solutions, when agitated with common air or oxygenous gas, appear luminous in the dark: a portion of the oil being rubbed upon the hand, makes it appear luminous.

The specific gravity of phosphorus is 1.7 nearly: the weight of its ultimate particle or atom is about 9 times that of hydrogen, as will appear when its compounds with oxygen are considered.

SECTION 7.

ON THE METALS.

The metals at present known, amount at least to 30 in number; they form a class of bodies which are remarkably distinguishable from others in several particulars, as well as from each other.

Gravity. One of the most striking properties of metals is their great weight or specific gravity. The lightest of them (excluding the lately discovered metals, potasium and sodium) weighs at least six times as much as water, and the heaviest of them 23 times as much. On the supposition that all aggregates are constituted of solid particles or atoms, each surrounded by an atmosphere of heat, it is a curious and important enquiry, whether this superior specific gravity of the metals is occasioned by the greater specific gravity of their individual solid particles, or from the greater number of them aggregated into a given volume, owing to some peculiar relation they may have to heat, or their superior attraction for each other. Upon examination of the facts exhibited by the metals, in their combinations with oxygen, sulphur, and the acids, it will appear that the former of these two positions is the true one ; namely, that the atoms of metals are heavier, almost in the same ratio as their specific gravities : thus an atom of lead will be found to be 11 or 12 times heavier than one of water, and its specific gravity is equally so. It must however be admitted, that in metals and other solid bodies, as well as in gases, their specific gravities are by no means *exactly* proportional to the weights of their atoms. It is further remarkable of the metals, that notwithstanding the great weight of their ultimate particles relatively, those particles possess no more, but often less, heat than particles of hydrogen, oxygen, or water. If the heat surrounding a particle of water of any temperature be denoted by 1, that surrounding a particle of lead will be found only $\frac{1}{2}$ as much, though the atom of lead is 12 times the weight of that of water. One would be apt to conclude from this circumstance, that an atom of lead has less attractive power for heat than an atom of water ; but this does not necessarily follow ; nay, the reverse is perhaps more pro bable of the two ; for, the absolute quantity of heat around any one particle in a state of aggregation, depends greatly upon the force of affinity, or the attraction of aggregation ;

if this be great, the heat is partly expressed or squeezed out; but if little, it is retained, though the attraction of the particles for heat remains unaltered. An atom of water may have the same attraction for heat that one of lead has; but the latter may have a stronger attraction of aggregation, by which a quantity of heat is expelled, and consequently less heat retained by any aggregate of the particles.

Opacity and Lustre. Metals are remarkably opake, or destitute of that property which glass and some other bodies possess, of transmitting light. When reduced to leaves as thin as possible, such as gold and silver leaf, they continue to obstruct the passage of light. Though the metallic atoms, with their atmospheres of heat, are nearly the same size as the atoms of water and their atmospheres, yet it seems highly probable that the metallic atoms abstracted from their atmospheres, are much larger than those of water in like circumstances. The former, I conceive, are large particles with highly condensed atmospheres; the latter, are small particles with more extensive atmospheres, because of their less powerful attraction for heat. Hence, it may be supposed, the opacity of metals and their lustre are occasioned. A great quantity of solid matter and a high condensation of heat, are likely to obstruct the passage of light, and to reflect it.

Malleability and Ductility. Metals are distinguished for these properties, which many of them possess in an eminent degree. By means of a hammer, they may be flattened and extended without losing their cohesion, especially if assisted by heat. Cylindrical rods of metal can be drawn through holes of less diameter, by which they are extended in length; and this successively till they form very small wire. These properties render them highly useful. Metals become harder and denser by being hammered.

Tenacity. Metals exceed most other bodies in their tenacity or force of cohesion; however they differ materially from each other in this respect. An iron wire of $\frac{1}{10}$th of an inch in diameter, will support 5 or 6 hundred pounds. Lead is only $\frac{1}{16}$th part as strong, and not equal to some sorts of wood.

Fusibility. Metals are fusible or capable of being melted by heat; but the temperatures at which they melt are extremely different.

Most of the metals possess a considerable degree of hardness; and some of them, as iron, are susceptible of a high degree of elasticity; they are mostly excellent conductors of heat and of electricity.

Metals combine with various portions of oxygen, and form metallic *oxides;* they also combine with sulphur, and form *sulphurets;* some of them with phosphorus, and form *phosphurets;* with *carbone* or charcoal, and form *carburets*, &c. which will be treated of in their respective places. Metals also form compounds one with another, called *alloys*.

The relative weights of the ultimate particles of the metals may be investigated, as will be shewn, from the metallic oxides, from the metallic sulphurets, or from the metallic salts; indeed, if the proportions of the several compounds can be accurately ascertained, I have no doubt they will all agree in assigning the same relative weight to the elementary particle of the same metal. In the present state of our knowledge, the results approximate to each other remarkably well, especially where the different compounds have been examined with care, and can be depended upon; but the proportions of the elements in some of the metallic oxides, sulphurets, and salts, have not yet been found with any degree of precision.

The number of metals hitherto discovered is 30, including the two derived from the fixed alkalis; some of these may, perhaps, be improperly denominated metals, as they are scarce, and have not been subjected to so much experience as others. The greater part of these metals have been discovered within the last century. Dr. Thomson divides the metals into 4 classes; 1. Malleable metals: 2. Brittle and easily fusible metals: 3. Brittle and difficultly fusible metals: 4. Refractory metals; that is, such as are known only in combination, it having not yet been found practicable to exhibit them in a separate state.—They may be arranged as follows:

1. Malleable.

1. Gold.	9. Copper.
2. Platinum.	10. Iron.
3. Silver.	11. Nickel.
4. Mercury.	12. Tin.
5. Palladium.	13. Lead.
6. Rhodium.	14. Zinc.
7. Iridium.	15. Potasium.
8. Osmium.	16. Sodium.

2. Brittle and easily fusible.

1. Bismuth.	3. Tellurium.
2. Antimony.	4. Arsenic.

3. Brittle and difficultly fused.

1. Cobalt.	4. Molybdenum.
2. Manganese.	5. Uranium.
3. Chromium.	6. Tungsten.

4. Refractory.

1. Titanium.
2. Columbium.
3. Tantalium.
4. Cerium.

To which last class also may the supposed metals from the earths be referred.

The following Table exhibits the chief properties of the metals in an absolute as well as comparative point of view.

Metals.	Colour.	Hardness.	Sp. Gr.	Wt. of ult. particles.	Melting point.	Tenacity.
Gold	yellow	6	19.362	140?	32° W	150
Platin.	white	7.5	23.00	100?	170°+W	274
Silver	white	6.5	10 511	100	22° W	187
Merc.	white	0.	13.580	167	—39° F	—
Pallad.	white	9.	11.871	—	160°+W	—
Rhod.	white	—	11.+	—	160°+W	—
Irid.	white	—	—	—	160°+W	—
Osm.	blue	—	—	—	160°+W	—
Copper	red	8	8.878	56	27° W	302
Iron	grey	9	7.788	50	158° W	550
Nickel	white	8	8.666	25? 50?	160°+W	—
Tin	white	6	7.300	50	410° F	31
Lead	blue	5	11.352	95	612° F	18
Zinc	white	6	7.190	56	680° F	18
Potas.	white	0.	.600	35	80° F	—
Sodium	white	1.	.935	21	150° F	—
Bism.	red wh	6	9.823	68?	476° F	20
Antim.	grey w.	6.5	6.860	40	810° F	7
Tellur.	blue w.	—	6.343	—	612°+F	—
Arsenic	blue w.	7	8.31	42?	400°+F	—
Cobalt	grey	8	7.811	55?	130° W	—
Mang.	grey	8	7.000	40?	160° W	—
Chrom.	yel. w.	—	—	—	170°+W	—
Uran.	iron gr.	—	9.000	60?	170°+W	—
Molyb.	yel. w.	—	7.500	—	170°+W	—
Tungs.	grey w.	9	17.6	56?	170°+W	—
Titan.	red	—	—	40?	170°+W	—
Colum.	—	—	—	—	170 +W	—
Tantal.	—	—	—	—	170 +W	—
Cerium	white	—	—	45?	170 +W	—

More particular Properties of the Metals.

Gold. This metal has been known from the earliest times, and always highly valued. Its scarcity, and several of its properties, contribute to make it a proper medium of exchange, which is one of its chief uses. English standard gold consists of 11 parts by weight of pure gold, and 1 part of copper (or silver) alloyed. This is usually spoken of as being 22 carats fine, pure gold being 24 carats fine. The use of the copper is to render the alloy harder, and consequently more durable than pure gold.

Gold retains its splendid yellow colour and lustre in all states of the atmosphere unchanged. Its specific gravity, when pure, and hammered, is 19.3, or more; but that of the same gold, in other circumstances, may be 19.2.—The specific gravity of standard gold varies from 17.1 to 17,9, accordingly as it is alloyed with copper, copper and silver, or silver, as well as from other circumstances. It excels all other metals in malleability and ductility; it may be beaten out so thin, that a leaf weighing 1 grain, shall cover 50 or 60 square inches, in which case the leaf is only $\frac{1}{280000}$th part of an inch in thickness: but it is capable of being reduced to $\frac{1}{12}$th of that thickness on silver wire. Gold melts at 32° of Wedgwood's pyrometer; that is, a red heat, but one greatly inferior to what may be obtained by a smith's forge: when fused, it may continue in that state for several weeks without losing any material weight. There is reason to believe that gold combines with oxygen, sulphur, and phosphorus; but those compounds are difficultly obtained. It combines with most of the metals, and forms alloys of various descriptions.

The weight of an atom of gold is not easily ascertained, because of the uncertainty in the proportions of the elements forming the compounds into which it enters. It is probably not less than 140, nor more than 200 times the weight of an atom of hydrogen.

PLATINA. This metal has not been found any where but in South America. In its crude state, it consists of small flattened grains of a metallic lustre, and a grey-white colour. This ore is found to be an alloy of several metals, of which platina is usually the most abundant. The grains are dissolved in nitro-muriatic acid, except a black matter which subsides; the clear liquor is decanted, and a solution of sal ammoniac is dropped into it: a yellow precipitate falls; this is heated to redness, and the powder is platina nearly pure. To obtain it still more pure, the process must be repeated upon this platina. When these grains are wrapped up in a thin plate of platina, heated to redness, and cautiously hammered, they unite and form a solid mass of malleable metal.

Platina thus obtained, is of a white colour, rather inferior to silver. In hardness it somewhat exceeds silver; but in specific gravity it exceeds all other bodies hitherto known. Specimens of it, when hammered, have been found of the specific gravity of 23 or upwards. It is nearly as ductile and malleable as gold. It requires a greater heat than most metals to fuse it; but when heated to whiteness, it welds in the same manner as iron. It is not in any degree altered by exposure to the air or to water. No ordinary artificial heat seems capable of burning it or uniting it to oxygen. Its oxidizement, however, may be effected by means of galvanism and electricity, and by exposing it to the heat excited by the combustion of hydrogen and charcoal in oxygenous gas. Platina has been united to phosphorus, but not to hydrogen, carbone, or sulphur. It unites with most of the metals to form alloys.

The weight of the ultimate particle of platina cannot be ascertained from the data we have at present: from its combination with oxygen, it should seem to be about 100; but, judging from its great specific gravity, one would be inclined to think it must be more. Indeed the proportion of oxygen in the oxides of platina cannot be considered as ascertained.

Platina is chiefly used for chemical purposes; in consequence of its infusibility, and the difficulty of oxidizing it, crucibles and other utensils are made of it, in preference to every other metal. Platina wires are extremely useful in electric and galvanic researches, for like reasons.

SILVER. This metal is found in various parts of the world, and in various combinations; but the greatest quantity is derived from America. Its uses are generally known. The specific gravity of melted silver is 10.474; after being hammered, 10.511. English standard silver, containing $\frac{1}{12}$ copper, simply fused, is 10.2. Pure silver is extremely malleable and ductile; but inferior in these respects to gold. It melts at a moderately red heat. It is not oxidized by exposure to the air, but is tarnished or loses its lustre, which is occasioned by the sulphureous vapours floating in the air. It unites with sulphur in a moderate heat; and may be oxidized by means of galvanism and electricity; it burns with a green flame.

Silver combines with phosphorus, and forms alloys with most of the metals.

The relative weight of an atom of silver admits of a pretty accurate approximation, from the known proportions of certain compounds into which silver enters ; namely, the oxides and sulphuret of silver, and the salts of silver : all of these nearly concur in determining the weight of an atom of silver to be 100 times that of hydrogen.

MERCURY. This metal, which is also known under the name of *quicksilver*, has been long discovered and in use. It is white and brilliant, reflecting more light from its surface, perhaps, than any other metal. Its specific gravity is 13.58. It is fluid at the common temperature of the atmosphere ; but it congeals when reduced to the temperature of — 39° Fahrenheit. It contracts suddenly at the point of congelation, contrary to what is exhibited in water ; when congealed, mercury becomes malleable ; but its qualities in a solid state are not easily to be ascertained. When heated in the open air to the temperature of 660°, or thereabouts, according to the equidifferential scale, mercury boils, and distils rapidly ; like water, however, it rises in vapour in a greater or less degree at all temperatures. Pure liquid mercury has no taste nor smell; it may be taken internally, without producing any remarkable effect on the human body. It can be united with oxygen, sulphur, and phosphorus; and it forms alloys, or, as they are more commonly called, *amalgams*, with most of the metals.

The weight of an atom of mercury is determinable from its oxides, its sulphuret, and the various salts which it forms with acids: from a comparison of all which, it seems to be about 167 times the weight of hydrogen. From any thing certainly known, the mercurial atom is heavier than any other; though there are two or three metals which exceed it in specific gravity.

PALLADIUM. This metal was discovered a few years ago in crude platina, by Dr. Wollaston, of which an account may be seen in the Philos. Transact. for 1804. It is a white metal, resembling platina in appearance, but is much harder: it is only one half of the specific gravity of platina. It requires great heat to fuse it, and is difficultly oxidized. Palladium combines with oxygen and sulphur, and forms alloys with several of the metals. But we have not yet sufficient data to determine the weight of its ultimate particles.

RHODIUM. This metal has been discovered still more recently than the last in crude platina, by Dr. Wollaston.—It constitutes about $\frac{1}{250}$th part of crude platina. It possesses nearly the same colour and specific gravity as palladium, and agrees with it in other particulars; but in certain respects they appear to possess essentially distinct properties.—The weight of the ultimate particles of this metal cannot yet be ascertained.

Iridium and Osmium. These two metals were lately discovered by Mr. Smithson Tennant to exist in crude platina. When crude platina is dissolved in nitro-muriatic acid, there remains a quantity of black shining powder; this powder contains two metals, one of which Mr. Tennant called *Iridium*, from the variety of colours which its solutions exhibited; the other *Osmium*, from a peculiar smell which accompanies its oxides. Iridium is a white metal, infusible as platina, difficultly soluble in any acid: it seems to combine with oxygen, and to form alloys with some of the metals. Osmium has a dark grey or blue colour: when heated in the air, it combines with oxygen, and the oxide is volatile, possessing the characteristic smell. In a close vessel, it resists any heat that has been applied; it also resists the action of acids, but unites with potash. It amalgamates with mercury. The weights of the atoms of these two metals are unknown.

Copper. This metal has been long known. It is of a fine red colour; its taste is styptic and nauseous. Its specific gravity varies from 8.6 to 8.9. It possesses great ductility, can be drawn into wire as fine as hair, and is capable of being beaten into very thin leaves. It is fused in a temperature higher than silver, and lower than gold, about 27° of Wedgwood's thermometer. Copper unites with oxygen, sulphur, and phosphorus; and forms alloys with several other metals.

The weight of the ultimate particle of copper, may be ascertained with considerable precision, from the proportions in which it is found combined with oxygen, sulphur, and phosphorus; as well as from its combinations with the acids. From a comparison of these, its weight seems to be nearly 56 times that of hydrogen.

Iron. This metal, the most useful we are acquainted with, has been long known. It seems to be found almost in every country, and in a great variety of combinations. Its ores require great heat to expel the foreign matters, and to melt the iron, which is first obtained in masses or pigs, called *cast iron*;

after which it undergoes a laborious operation, the object of which is to expel the carbone and oxygen which it may yet contain, and to render it malleable. This consists chiefly in hammering the iron when heated almost to fusion.

Iron is susceptible of a high polish ; it is very hard ; it varies in specific gravity from 7.6 to 7.8. It is distinguishable from all other metals, by possessing, in a high degree, (indeed almost exclusively) magnetical attraction. The magnet or loadstone itself is chiefly iron, with certain modifications. Iron increases in malleability as it increases in temperature : its ductility is surpassed by few other metals, as its wire admits of extension till it becomes as fine as human hair : its tenacity, which is one of its most valuable properties, is not equalled by any other body we are acquainted with. Pure malleable iron is estimated to melt at 158° of Wedgwood ; whereas cast iron melts about 130°.

Iron is distinguished for its combinations with oxygen, carbone, sulphur, and phosphorus : it forms alloys with several of the metals, but they are not of much importance.

The weight of an atom of iron may be found from almost any of its numerous combinations, either its oxides, its sulphurets, or any of the salts which it forms with acids : all these will be found to give the same weight nearly ; namely, 50 times the weight of an atom of hydrogen.

NICKEL. The ore from which this metal is obtained, is found in Germany : it usually contains several other metals, from which it is difficult to extract the nickel in a state of tolerable purity. Nickel, when pure as it can be obtained, is of a silver white colour ; its specific gravity is 8.279, and when forged 8.666. It is malleable, both hot and cold, and may be beaten into a leaf of $\frac{1}{100}$ of an inch in thickness. A very great heat is required to fuse it. It is attracted by the magnet nearly as much as iron, and may be converted into a magnet itself. It combines with oxygen, sulphur, and phosphorus ; and may be alloyed with certain other metals.

The weight of its atom can scarcely yet be determined, for want of a more accurate knowledge of the compounds into which it enters : perhaps it will be found to weigh about 25, or else double that number, 50.

TIN. This metal has been long known, though it is found but in few places comparatively. Cornwall is the only part of Great Britain where this metal abounds ; and its tin mines are the most celebrated in Europe. Tin is a white metal, nearly resembling silver; its specific gravity is about 7.3. It is malleable in a high degree; but inferior to many metals in ductility and tenacity. It melts at the low temperature of 440° Fahrenheit. When exposed to the air, it loses its lustre, and becomes grey; this is more rapidly the case if it be melted; its surface then soon becomes grey, and in time passes to yellow Tin combines with oxygen, sulphur, and phosphorus, and forms alloys with most of the metals.

The weight of an atom of tin may be derived from the proportion of the elements in the oxides, the sulphuret or the phosphuret of tin; or from the salts of tin. It is probably about 50 times heavier than hydrogen.

Lead. This metal seems to have been known in early times: it is of a blueish white colour, bright when recently melted, but soon loses its lustre when exposed to the air. It has scarcely any taste or smell; but operates as a deadly poison when taken internally: it seems to benumb the vital functions, and to destroy the nervous sensibility, inducing a paralysis, and finally death. The specific gravity of lead, whether hammered or not, is about 11.3 or 11.4; it is malleable, and may be reduced to thin plates. It melts about 610° of Fahrenheit. It combines with oxygen, sulphur, and phosphorus, and forms alloys with most other metals.

The ultimate particle of lead, as deduced from a comparison of its oxides, sulphuret, and the salts in which it is found, I estimate at 95 times that of hydrogen.

ZINC. The ores of this metal are not rare; but the metal has not been extracted from them in a pure state, at least in Britain, much more than half a century. Zinc is a brilliant white metal, inclining to blue. Its specific gravity is from 6.9 to 7.2. It was till lately considered as a brittle metal; but Messrs. Hobson and Sylvester, of Sheffield, have discovered that between the temperature of 210° and 300°, zinc yields to the hammer, may be laminated, wire drawn, &c. and that after being thus wrought, it continues soft and flexible. It melts about 680°, and above that temperature evaporates considerably. Zinc soon loses its lustre in the air, and grows grey; but in water it becomes black, and hydrogen gas is emitted. Zinc combines with oxygen; and either it or its oxides combine with sulphur and phosphorus. It forms alloys with most of the metals, some of which are very useful.

The atom of zinc weighs nearly 56 times as much as hydrogen.

POTASIUM. We are principally indebted to

Mr. Davy for our knowledge of this metal; its oxide, potash, or the fixed vegetable alkali, is universally known; but the decomposition of the oxide is a recent discovery. To obtain the metal, a small piece (30 or 40 grains) of pure caustic potash, which has been exposed to the air a few moments, to acquire a slight degree of moisture, sufficient to render it a conductor of galvanism, is to be exposed to the action of a powerful galvanic battery; by its operation, the oxygen of the potash is expelled, and fluid metallic globules of the appearance of mercury, are obtained. This metal has also been produced by Messrs. Gay Lussac and Thenard, by exposing potash to iron turnings in a white heat: some potasium was obtained, and an alloy of potasium and iron. Mr. Davy has made an experiment with a similar result; and found that a large quantity of hydrogen gas is at the same time given out. This fact seems to point out potash as a compound of potasium and water, and not of potasium and oxygen; the French chemists argue that potasium is a compound of hydrogen and potash; but, as Mr. Davy properly observes, their argument amounts to this, that potasium is a compound of hydrogen and an unknown base, which compound united to oxygen forms potash. This subject must be left to future experience.—Potasium, at the temperature of 32°, is solid and brittle; and its fragments exhibit a crystallized texture: at 50°, it is soft and malleable; at 60°, it is imperfectly fluid; at 100°, it is perfectly fluid, and small globules unite as in mercury. It may be distilled by a heat approaching to redness. I s specific gravity is only 6; this circumstance would seem to countenance the notion of its containing hydrogen. Potasium combines with oxygen, sulphur, and phosphorus; and it seems to form alloys with many of the metals.

The weight of an atom of potasium appears from its combination with oxygen to be 35 times that of hydrogen.

SODIUM. Mr. Davy obtained this metal from the fixed mineral alkali, or soda, by means of galvanism, in the same way as potasium. Sodium, at the common temperature, is a solid, white metal, having the appearance of silver; it is exceedingly malleable, and much softer than other metallic substances. Its specific gravity is rather less than water, being 9348. It begins to melt at 120°, and is perfectly fluid at 180°. It combines with oxygen, sulphur, and phosphorus; and forms alloys with the metals.

The weight of an atom of sodium, as deduced from its combination with oxygen, is nearly 21 times the weight of hydrogen.

Bismuth. This has not been known as a distinct metal much more than a century. Its ores are found chiefly in Germany.—Bismuth is of a reddish white colour; it loses its lustre by exposure to the air; its specific gravity is about 9.8; it is hard, but breaks with a smart stroke of a hammer; it melts about 480°. In a strong red heat, bismuth burns with a blue flame, and emits yellow fumes. It combines with oxygen and sulphur, and forms alloys with most of the metals.

The weight of an atom of bismuth, may be derived from its oxides and sulphuret: it seems to be about 68 times the weight of an atom of hydrogen.

Antimony. Some of the ores of this metal were known to the ancients; but the metal in a pure state, has not been known more than 300 years. Antimony has a greyish white colour, and considerable brilliancy; its specific gravity is 6.7 or 6.8; it is very brittle; it melts about 810° Fahrenheit; it loses its lustre in time by exposure to the air. Antimony combines with oxygen, sulphur, and phosphorus; and it forms alloys with most of the other metals.

The weight of an atom of antimony, is

determinable from its compounds with oxygen and sulphur, and seems to be 40 times the weight of hydrogen.

ARSENIC. Certain compounds of Arsenic were known to the ancients. It seems to have been known in a distinct character for more than a century. Arsenic has a blueish grey colour, and considerable brilliancy, which it soon loses by exposure to the air; its specific gravity is stated to be 8.3; its fusing point has not been ascertained, by reason of its great volatility: it has been heated to 350°, at which temperature it sublimes quickly, and exhibits a strong smell resembling that of garlic, which is characteristic of this metal. It combines with oxygen, forming one of the most virulent poisons; also with hydrogen, sulphur, and phosphorus; and it forms alloys with most of the metals.

The weight of an atom of arsenic, appears from its compounds to be 42 times that of hydrogen.

COBALT. The ore of this metal has been long used to tinge glass blue; but it was not till the last century that a peculiar metal was extracted from it. Cobalt is of a grey colour, inclining to red; it has not much lustre: its specific gravity is about 7.8; it is brittle; it melts at 180° of Wedgwood; it is attracted by the magnet, and is itself capable of being made magnetic, according to Wenzel. Cobalt combines with oxygen, sulphur, and phosphorus; and it forms alloys with most of the metals, but they are of little importance.

The weight of an atom of cobalt cannot be accurately obtained from the data we have at present; it is probably 50 or 60 times that of hydrogen.

MANGANESE. The dark brown mineral called manganese, has been known and used in the glass manufactories, perhaps more than a century: but the metal which now goes by the same name, was not discovered till about 40 years ago: in fact, it is not yet much known, being obtained with difficulty, and by a great heat. The metal is of a greyish white colour, and considerable brilliancy: its specific gravity is 6.85 or 7; it is brittle, and melts at 160° of Wedgwood; when reduced to powder it is attracted by the magnet, which is supposed to be owing to the presence of iron. Manganese attracts oxygen from the air, becoming grey, brown, and finally black. It is capable of being combined with sulphur and phosphorus; and it forms alloys with some of the metals, but they have not been much examined.

The weight of an atom of manganese, as

determined from its oxides, seems to be about 40 times that of hydrogen.

CHROMIUM. This metal, united to oxygen so as to constitute an acid, is found in the *red lead ore* of Siberia. The pure metal being obtained, is white inclining to yellow; it is brittle, and requires a great heat to fuse it. It combines with oxygen. The other properties of this metal are not yet known. Its atom, perhaps, weighs about 12 times that of hydrogen.

URANIUM. This metal was discovered by Klaproth, in 1789, in a mineral found in Saxony. It is obtained with some difficulty, and only in small quantities; it has, therefore, been examined but by few. The colour of uranium is iron grey; it has considerable lustre; it yields to the file; its specific gravity is 8.1, according to Klaproth; 9.0, according to Bucholz. Uranium unites with oxygen, and probably with sulphur: its alloys have not been ascertained.

The weight of an atom of this metal, is probably about 60 times that of hydrogen.

MOLYBDENUM. The ore from which this metal is obtained is a sulphuret, called *molybdena*; but it requires an extraordinary heat to reduce it; the metal has not hitherto been obtained, except in small grains. It is of a yellowish white colour; its specific gravity is 7.4, according to Hielm; but 8.6, according to Bucholz. It combines with oxygen, sulphur, and phosphorus; and it forms alloys with several of the metals.

The atom of molybdenum, probably weighs about 60 times that of hydrogen.

TUNGSTEN. This metal is one of those recently discovered. It is difficultly obtained, requiring an excessive heat for its fusion. It is of a greyish white colour, and considerable brilliancy; its specific gravity is 17.2 or 17.6; it is very hard, being scarcely impressed with a file. It combines with oxygen, sulphur, and phosphorus; and it forms alloys with other metals.

We have not sufficient data, from which to determine the weight of an atom of tungsten: as far as we can judge from its oxides, its weight must be 55 times that of hydrogen, or upwards.

TITANIUM. This metal has been lately discovered. It is said to be of a dark copper colour; it has much brilliancy, is brittle, and possesses in small scales a considerable degree of elasticity. It is highly infusible. It tarnishes on exposure to the air; is oxidized by heat, and then becomes blueish. It unites with phosphorus, and has been alloyed with iron. It detonates when thrown into red hot nitre. The atom of titanium probably weighs about 40 or 50 times that of hydrogen.

Columbium. In 1802, Mr. Hatchett discovered a new metallic acid in an ore containing iron, from America. He did not succeed in reducing the acid to a metal; but, from the phenomena it exhibited, there was little room to doubt of its containing a peculiar metal, which he called columbium.

Tantalium. This metal has lately been discovered by M. Ekeberg, a Swedish chemist. A white powder is extracted from certain minerals, which appears to be an oxide of this metal. When this white oxide is strongly heated along with charcoal, in a crucible, a metallic button is formed, of external lustre, but black and void of lustre within. The acids again convert it into the state of a white oxide, which does not alter its colour when heated to redness.

Cerium. The oxide of this metal is obtained from a Swedish mineral. No one has yet succeeded completely in reducing this oxide; so that the properties of the metal, and even its existence, are yet unknown. But the earth or supposed oxide, is found to have properties similar to those of other oxides. These, of course, belong to a future article, the metallic oxides.

CHAP. V.

COMPOUNDS OF TWO ELEMENTS.

IN order to understand what is intended to be signified by *binary* and *ternary* compounds, &c. the reader is referred to page 213 and seq. Some persons are used to denominate all compounds, where only two elements can be discovered, *binary* compounds; such, for instance, as nitrous gas, nitrous oxide, nitric acid, &c. in all of which we find only azote and oxygen. But it is more consistent with our views to restrict the term *binary*, to signify *two* atoms; *ternary*, to signify *three* atoms, &c. whether those atoms be elementary or otherwise; that is, whether they are the atoms of undecompounded bodies, as hydrogen and oxygen, or the atoms of compound bodies, as water and ammonia.

In each of the following sections, we shall consider the compounds of some two of the elementary or undecompounded bodies; beginning each section with the *binary* compounds, then proceeding to the *ternary* compounds, or at least to those which consist of *three* atoms, though they may be *binary* in the sense we use the term; and so on to the more complex forms.

This chapter will comprehend all the aeriform bodies that have not been considered in the last, several of the acids, the alkalies, the earths, and the metallic oxides, sulphurets, carburets, and phosphurets.

In treating of these articles, I intend to adopt the most common names for them; but it will be obvious, that if the doctrine herein contained be established, a renovation of the chemical nomenclature will in some cases be expedient.

SECTION 1.

OXYGEN WITH HYDRONEN.

1. *Water.*

This liquid, the most useful and abundant of any in nature, is now well known both by analytic and synthetic methods, to be a compound of the two elements, oxygen and hydrogen.

Canton has proved that water is in degree compressible. The expansive effect of heat on water has been already pointed out. The weight of a cubic foot of water is very near 1000 ounces avoirdupoise. This fluid is commonly taken as the standard for comparing the specific gravities of bodies, its weight being denoted by unity.

Distilled water is the purest; next to that, rain water; then river water; and, lastly, spring water. By purity in this place, is meant freedom from any foreign body in a state of solution; but in regard to transparency, and an agreeable taste, spring water generally excels the others. Pure water has the quality we call *soft*; spring and other impure water has the quality we call *hard*. Every one knows the great difference of waters in these respects; yet it is seldom that the hardest spring water contains so much as $\frac{1}{1000}$th part of its weight of any foreign body in solution. The substances held in solution are usually carbonate and sulphate of lime.

Water usually contains about 2 per cent. of its bulk of common air. This air is originally forced into it by the pressure of the atmosphere; and can be expelled again no other way than by removing that pressure. This may be done by an air-pump; or it may in great part be effected by subjecting the water to ebullition, in which case steam takes the place of the incumbent air, and its pressure is found inadequate to restrain the dilatation of the air in the water, which of course makes its escape. But it is difficult to expel all the air by either of those operations. Air expelled from common spring water, after losing 5 or 10 per cent. of carbonic acid, consists of 38 per cent. of oxygen and 62 of azote.

Water is distinguished for entering into combination with other bodies. To some it unites in a small definite proportion, constituting a solid compound. This is the case in its combination with the fixed alkalies, lime, and with a great number of salts; the compounds are either dry powders or crystals. Such compounds have received the name of *hydrates*. But when the water is in excess, a different sort of combination seems to take place, which is called *solution*. In this case, the compound is *liquid* and transparent; as when common salt or sugar are dissolved in water. When any body is thus dissolved in water, it may be uniformly diffused through any larger quantity of that liquid, and seems to continue so, without manifesting any tendency to subside, as far as is known.

In 1781, the composition and decomposition of water were ascertained; the former by Watt and Cavendish, and the latter by Lavoisier and Meusnier. The first experiment on the composition of water on a large scale, was made by Monge, in 1783; he procured about $\frac{1}{4}$ lb. of water, by the combustion of hydrogen gas, and noted the quantities of hydrogen and oxygen gas which had disappeared. The second experiment was made by Le Fevre de Gineau, in 1788; he obtained about $2\frac{1}{2}$ lbs. of water in the same way. The third was made by Fourcroy, Vauquelin, and Seguin, in 1790, in which more than a pound of water was obtained. The general result was, that 85 parts by weight of oxygen unite to 15 of hydrogen to form 100 parts of water. —Experiments to ascertain the proportion of the elements arising from the decomposition of water, were made by Le Fevre de Gineau and by Lavoisier, by transmitting steam through a red hot tube containing a quantity of soft iron wire; the oxygen of the water combined with the iron, and the hydrogen was collected in gas. The same proportion, or 85 parts of oxygen and 15 of hydrogen, were found as in the composition.

The Dutch chemists, Dieman and Troostwyk, first succeeded in decomposing water by electricity, in 1789. The effect is now produced readily by galvanism. The composition of water is easily and elegantly shewn, by means of Volta's eudiometer, an instrument of the greatest importance in researches concerning elastic fluids. It consists of a strong graduated glass tube, into which a wire is hermetically sealed, or strongly cemented ; another detached wire is pushed up the tube, nearly to meet the former, so that an electric spark or shock can be sent from one wire to the other through any portion of gas, or mixture of gases, confined by water or mercury. The end of the tube being immersed in a liquid, when an explosion takes place, no communication with the external air can arise; so that the change produced is capable of being ascertained.

The component parts of water being clearly established, it becomes of importance to determine with as much precision as possible, the relative weights of the two elements constituting that liquid. The mean results of analysis and synthesis, have given 85 parts of oxygen and 15 of hydrogen, which are generally adopted. In this estimate, I think, the quantity of hydrogen is overrated. There is an excellent memoir in the 53d vol. of the Annal. de Chemie, 1805, by Humboldtd and Gay-Lussac, on the proportion of oxygen and hydrogen in water. They make it appear, that the quantity of aqueous vapour which elastic fluids usually contain, will so far influence the weight of hydrogen gas, as to change the more accurate result of Fourcroy, &c. of 85.7 oxygen and 14.3 hydrogen, to 87.4 oxygen and 12.6 hydrogen. Their reasoning appears to me perfectly satisfactory. The relation of these two numbers is that of 7 to 1 nearly. There is another consideration which seems to put this matter beyond doubt. In Volta's eudiometer, *two* measures of hydrogen require just *one* of oxygen to saturate them. Now, the accurate experiments of Cavendish and Lavoisier, have shewn that oxygen is nearly 14 times the weight of hydrogen; the exact coincidence of this with the conclusion above deduced, is a sufficient confirmation.—If, however, any one chooses to adopt the common estimate of 85 to 15, then the relation of oxygen to hydrogen will be as $5\frac{2}{3}$ to 1; this would require the weight of oxygenous gas to be only $11\frac{1}{3}$ times the weight of hydrogen.

The absolute weights of oxygen and hydrogen in water being determined, the relative weights of their atoms may be investigated. As only *one* compound of oxygen and hydrogen is certainly known, it is agreeable to the 1st rule, page 214, that water should be concluded a *binary* compound; or, one atom of oxygen unites with one of hydrogen to form one of water. Hence, the relative weights of the atoms of oxygen and hydrogen are 7 to 1.

The above conclusion is strongly corroborated by other considerations. Whatever may be the proportions in which oxygen and hydrogen are mixed, whether 20 measures of oxygen to 2 of hydrogen, or 20 of hydrogen to 2 of oxygen, still when an electric spark is passed, water is formed by the union of 2 measures of hydrogen with 1 of oxygen, and the surplus gas is unchanged. Again, when water is decomposed by electricity, or by other agents, no other elements than oxygen and hydrogen are obtained. Besides, all the other compounds into which those two elements enter, will in the sequel be found to support the same conclusion.

After all, it must be allowed to be possible that water may be a ternary compound. In this case, if two atoms of hydrogen unite to one of oxygen, then an atom of oxygen must weigh 14 times as much as one of hydrogen; if two atoms of oxygen unite to one of hydrogen, then an atom of oxygen must weigh $3\frac{1}{2}$ times one of hydrogen.

2. *Fluoric Acid.*

The acid obtained from the fluor spar, which abounds in Derbyshire, is one of those the base of which has not yet been clearly ascertained; but, guided partly by theoretic reasoning, and partly by experience, I have ventured to place it among the compounds of hydrogen with oxygen, and to rank it next to water in simplicity of constitution; it is, as I conceive, a compound of two atoms of oxygen with one of hydrogen.

Scheele and Priestley have distinguished themselves in investigating the properties of this acid; and Dr. Henry and Mr. Davy have attempted to decompose it. The acid may be obtained by taking a quantity of pounded fluor spar (fluate of lime), putting it into a gas bottle with about the same weight of sulphuric acid undiluted, and then applying a heat, so as to raise the temperature to about the boiling heat of water. The acid is produced in the gaseous form, and must be received over mercury; but if it is intended to condense it in water, then the gas, as it is generated, may be sent into a receiver containing some water at the bottom; the water will rapidly absorb the gas, and increase in density.

Some of the properties of this acid are, 1. In the elastic state it is destructive of combustion, and of animal life; it has a pungent smell, somewhat like muriatic acid, and not less suffocating; its specific gravity has not been accurately obtained; but from some experiments I have made, it seems to be extremely heavy when obtained in glass vessels; in fact, it is in that case a superfluate of silica: Into a clean dry flask, I sent a quantity of fluoric acid gas; after some time, the mixture of common air and acid was corked, and the flask weighed: it had acquired 12 grains. The flask was next inverted in water, to see how much would be absorbed, and that quantity was taken for the acid gas. The capacity of the flask was 26 cubic inches, containing originally 8.2 grains of common air; 12 cubic inches of acid gas had entered. According to this, if the whole flask had been filled with the gas, it would have gained 26 grains; consequently, 26 cubic inches of the acid gas would weigh 34.2 grains, and its specific gravity be 4.17 times that of common air. This experiment was repeated with a proportional result. The flask became partially lined with a thin, dry film of fluate of silica during the operation, which no doubt contributed something to the weight; but I am convinced, from other experiments, that this gas, when loaden with silica, is heavier than most others. A tube, four tenths of an inch in diameter, and 10 inches long, being filled with this acid gas, and inverted for one minute, retained only $\frac{35}{220}$ths of the gas; whereas, with carbonic acid gas, it retained $\frac{140}{220}$ths; and with oxymuriatic acid gas, $\frac{65}{220}$ths. 2. Water absorbs a very large portion of this gas; but the quantity is, like as in other similar cases, regulated by the temperature and pressure conjointly: at the common temperature and pressure, I have observed 2 grains of water take up 200 times their bulk of the gas, and leave little residuum besides common air. It is seldom obtained in large quantities of this strength; when water has imbibed its bulk of the gas, it has a sour taste, and all the other characters of acids. 3. The property of dissolving silica (flint) is peculiar to this acid; when it is obtained, as usual, in glass vessels, it corrodes the glass, and takes up a portion of silica, which is held in solution in the transparent gas; but as soon as this comes in contact with water, the silica is deposited in form of a white crust, namely, fluate of silica, on the surface of the water. 4. The gas, when thrown into common air, exhibits white fumes (like muriatic acid); this is owing to its combining with the steam or aqueous vapour, which common air always contains in a diffused state. 5. Fluoric acid combines with the alkalies, earths, and metallic oxides, forming salts denominated *fluates*.

The weight of an atom of fluoric acid may be investigated from the salts into which it enters as an integral element. Of these, the *fluate of lime* is most abundant, and best known. Scheele is said to have found 57 parts of lime, and 43 of acid and water, in fluate of lime. Richter finds 65 lime, and 35 acid in this salt. These are the only authorities I know: they differ materially. In order to satisfy myself, I took 50 grains of finely pulverized spar, and having mixed with it as much, or more, strong sulphuric acid, the whole was exposed to a heat gradually increasing to redness; the result was, a hard dry crust of mixed sulphate and fluate of lime; this was pulverized, then weighed, and again mixed with sulphuric acid, and heated as before; this process was repeated two or three times, or as long as any increase of weight was found. At last, a dry white powder, of 75 grains, was obtained, which was pure sulphate of lime. This experiment, two or three times repeated, gave always 75 grains finally. Hence, 50 grains of fluate of lime contain just as much lime as 75 grains of sulphate of lime: But sulphate of lime is formed of 34 parts acid + 23 parts lime; now, if 57 : 23 : : 75 : 30 = the lime in 50 fluate of lime. Hence, fluate of lime consists of 60 lime + 40 acid, in 100 parts: a result which is nearly a mean between the two beforementioned. Again, if 60 : 40 : : 23 : 15 nearly, for the weight of fluoric acid which is found associated with 23 parts of lime; but 23 will be found in the sequel to represent the weight of an atom of lime; therefore, 15 represents the weight of an atom of fluoric acid, it being assumed that fluate of lime is constituted of one atom of acid united to one atom of lime.

Before we commence the analytical investigation of this acid, it will be proper to discuss its relation to steam or aqueous vapour, which appears at present to be much misunderstood; the observations equally apply to muriatic acid gas, and to some others, which will be noticed in their places. It is universally known, that common air over water contains a quantity of steam or vapour, some way or other combined or mixed with it, which does not impair its transparency, but which gives it $\frac{1}{50}$th of its elastic force, at the temperature of 65°; the vapour too, increases and diminishes in force and quantity in same ratio with the temperature. Clement and Desormes have shewn, that this vapour is the same in quantity for atmospheric air, oxygen, hydrogen, azote, and carbonic acid, and probably for most other gases. This vapour can be abstracted from the gases by any body possessing an attraction for water; such as sulphuric acid, lime, &c. In short, it can be taken out, as far as is known, by any body that will take out pure steam. Some authors consider the vapour united to the air by a slight affinity; others call it hygrometrical affinity, &c. My opinion on this subject has already been stated, that the steam mixed with air differs in no respect from pure steam; and, consequently, is subject to the same laws. There are some elastic fluids, however, which have so strong an affinity for water, that they will not permit this steam quietly to associate with them; these are fluoric, muriatic, sulphuric, and nitric acids. No sooner are these acid gases presented to any air containing steam, but they seize upon the steam; the two united, are converted into a liquid; visible fumes appear, which after playing about a while, are observed to fall down, or adhere to the sides of the vessel, till the gas no longer finding any steam present, occupies the volume of the vessel in a transparent state, free from every atom of vapour. These acid gases cannot exist one moment along with steam; they are no longer elastic fluids, but liquids; the drops of liquid float about, and cause the visibility, till, like rain, they subside; they are not reabsorbed; for, if the surface of a glass vessel is once moistened with them, it remains so. Hence, it should seem that these acid gases, so far from obstinately retaining their vapour, as is commonly imagined, they cannot be induced to admit any vapour at all, in ordinary circumstances. This being clearly understood, we can now proceed to consider the experiments on the analysis of fluoric acid.

In the Philos. Transact. for 1800, Dr. Henry has given us an interesting set of experiments on the decomposition of the muriatic acid by electricity: at the conclusion, he observes on fluoric acid—" When electrified alone, in a " glass tube, coated internally with wax, it " sustained a diminution of bulk, and there " remained a portion of hydrogenous gas." Now, admitting the accuracy of the fact, it seems fair to infer, that hydrogen is a constituent principle of fluoric acid; and not, as he supposed, derived from the water it contains. More recently, Mr. Davy has ascertained, (see Philos. Transact. for 1808) that potasium burns in fluoric acid, and the result is fluate of potash, and a little hydrogen gas is liberated. In particular, $10\frac{1}{2}$ grains of potasium were burned in 19 cubic inches of fluoric acid, 14 of which disappeared, fluate of potash was formed, and $2\frac{1}{4}$ cubic inches of hydrogen were evolved. Here it is evident, that both oxygen and hydrogen were found in the fluoric acid, and must have made an integral part of that acid, as no vapour could subsist in it; whence it appears, that both oxygen and hydrogen are essential to fluoric acid. Moreover, it is highly probable that the pure acid in the 14 inches of gas, weighed about 6 grains, (common air being $4\frac{1}{2}$) and the oxygen necessary for $10\frac{1}{2}$ potasium, would be 2 grains; whence the acid entering into composition, would be about twice the weight of the oxygen united to the potasium.

I shall now relate some of my own experiments on the decomposition of this acid.

1. Fluoric acid gas may, I find, be kept in glass tubes for several hours or days, without any change of bulk; it continues at the end absorbable by water as at first. Two successive trials were made, by electrifying about 30 water grain measures of the gas. After two hours electrification, no change of volume was produced. Water was then admitted, which absorbed all but 4 grain measures; to this 14 measures of hydrogen were added, and

a sufficient quantity of oxygen ; the whole was then exploded, and a diminution of 23.3 was observed, denoting 15.5 hydrogen. Here seems, then, to have been a decomposition of the acid, and a formation of 1.5 hydrogen. This was the result of the latter experiment, and the former was to the same effect.

2. Fluoric acid gas, electrified along with hydrogen, experiences a diminution, but this is much greater in the hydrogen than in the acid. The result of one of the most careful experiments follows. A mixture of 20 measures of fluoric acid, and 13 of hydrogen, was electrified for three hours uninterruptedly, by a dense stream of sparks; it diminished from 33 to 19 ; of the loss, 10 was found to be hydrogen, and 4 acid.—Here the hydrogen must, probably, have formed water with part of the oxygen of the acid.

3. Fluoric acid was mixed with oxygen, and electrified one hour ; a small diminution was observed, and the surface of the mercury was tarnished.

4. Fluoric acid gas was mixed with oxymuriatic acid gas : no sensible change was produced.

Upon the whole, it appears that the weight of an atom of fluoric acid is about 15 times that of hydrogen, that it contains hydrogen and oxygen, and nothing besides as far as is certainly known. Now, as the weight of one atom of hydrogen, and two of oxygen, just make 15 times that of hydrogen, there is great reason to presume that this must be the constitution of that acid. Besides, analogy is strongly in favour of the conclusion; an atom of the other elementary principles, azote, carbone, sulphur, and phosphorus, joined to two atoms of oxygen, each forms a peculiar acid, as will be shewn in the sequel; why, then, should not one atom of hydrogen and two of oxygen, also form an acid?

3. *Muriatic Acid.*

To obtain muriatic acid in the elastic state, a portion of common salt, muriate of soda, is put into a gas bottle, and about an equal weight of concentrated sulphuric acid; by the application of a moderate heat to the mixture, a gas comes over, which may be exhibited over mercury; it is muriatic acid gas.

Some of the properties of muriatic acid gas, are: 1. It is an invisible elastic fluid, having a pungent smell; it is unfit for respiration, or for the support of combustion; when mixed with common air, it produces a white cloud, which is owing to its combination with steam, and the consequent formation of innumerable small drops of liquid muriatic acid. 2. Its specific gravity appears to be about 1.61 times that of common air, from some experiments of mine; but, according to Brisson, it is 1.43; and according to Kirwan, 1.93 at the temperature of 60°, and pressure of 30 inches of mercury. There are two sources of error obvious in determining its specific gravity; the one is, that liquid muriatic acid is apt to insinuate itself, if the utmost attention is not paid to have the mercury in the vessel dry, in which case the weight is found too great; this is probably Kirwan's error: the other is, a quantity of common air may be mixed with the acid gas, in which case its weight will be too little. In order to find the specific gravity of this gas, I adopted the same method as with fluoric acid (see page 278). A flask containing 8.2 grains of common air, when partially filled with muriatic acid gas, (namely $\frac{17}{28}$ths) acquired just 3 grains; and a like proportion in several other trials; from which I find the specific gravity given above. 3. It possesses the characterisic properties of acids; namely, that of converting vegetable blues to red, of uniting with alkalies, &c. 4. It is rapidly and largely absorbed by water, which takes up between four and five hundred times its bulk of the gas, at the common temperature and pressure; that is, rather less than an equal weight. This combination of water and muriatic acid gas, constitutes the common liquid muriatic acid, or spirit of salt of commerce; but it is never of the strength indicated above. It is usually of a yellow colour, owing to some atoms of iron which it holds in solution.

The constitution of this acid, is a question that has long engaged the attention of chemists. This acid seems more difficultly decomposed than most others. Electricity, so powerful an agent in the composition and decomposition of other acids, seems to fail in this. In the Phil. Tr. for 1800, Dr. Henry has given us the results of a laborious investigation on this subject. From these it appears that pure, dry muriatic acid gas, is scarcely affected by electricity. A very small diminution in volume, and some traces of hydrogenous gas, were observed, which he ascribes to the water or steam which the gas contains. But we have already remarked, (page 283) that muriatic acid gas naturally contains no steam; or, if it contains any, it must be much less than other gases contain. It is probable, therefore, that the hydrogen was derived from the decomposition of part of the acid. This conclusion is strengthened by the recent experiments of Mr. Davy, in which the acid has apparently undergone a complete decomposition. In his Electrochemical Researches, in the Philos. Transact. for 1808, he observes—' When potasium was heated in muriatic acid gas, as dry as it could be obtained ' by common chemical means, there was a ' violent chemical action with ignition; and ' when the potasium was in sufficient quantity, the muriatic acid gas wholly disappeared, and from one third to one fourth of ' its volume of hydrogene was evolved, and ' muriate of potash was formed.' Here it is almost certain a portion of the acid was decomposed; the residuary hydrogen, and the oxygen required to convert the potasium into potash, are the only ostensible elements of the acid; hence we must infer, that muriatic acid is a compound of oxygen and hydrogen. In a subsequent paper in the same volume, Mr. Davy informs us, that 8 grains of potasium took 22 cubic inches of acid gas, and gave 8 inches of hydrogen. This particular experiment must, however, be incorrect in some point; or otherwise the general observation before made; because they are inconsistent with each other. For, 22 cubic inches of acid gas weigh 11 grains, to which adding 8 grains of potasium, we obtain 19 grains; but 8 grains of potasium form only 14.6 grains of muriate of potash, to which adding .2 grain for the 8 cubic inches of hydrogen, gives 14.8 instead of 19 grains. I would therefore adopt the general fact, which was confirmed by several experiments, and is entirely consistent; namely, that *when potasium in sufficient quantity is burned in muriatic acid gas, the whole of the gas disappears, and from one third to one fourth of its volume of hydrogen is evolved, and muriate of potash formed.* This is one of the most important facts that has been ascertained, respecting the constitution of muriatic acid. Now, the elements of muriate of potash are as follow: 35 grains of potasium + 7 of oxygen = 42 of potash; and 42 potash + 22 muriatic acid = 64 grains of muriate of potash. From this it appears, that the oxygen in muriate of potash is nearly $\frac{1}{3}$ of the weight of the acid. According to this, when potasium is burned in muriatic acid gas, nearly $\frac{1}{4}$ of the whole weight (for the hydrogen weighs little) goes to the oxidizement of the potasium, and the remaining $\frac{3}{4}$ unite with the potash formed. Hence, when 22 cubic inches, or 11 grains of gas disappear, as in the particular experiment lately mentioned, $2\frac{3}{4}$ grains nearly must have been oxygen derived from the acid, and $8\frac{1}{4}$ grains of acid joined to the potash so formed. But $2\frac{3}{4}$ grains of oxygen = 8 cubic inches, would require 16 inches of hydrogen to form water: it is evident, then, that water was not the source of the oxygen; for, if it had, there must have been twice the quantity of hydrogen evolved. Mr. Davy has ascertained another fact, exactly similar to the general one just stated; namely, that when charcoal is galvanized in muriatic acid gas, muriate of mercury is formed, and hydrogen, amounting to $\frac{1}{3}$ of the volume of the gas is evolved. He infers from this, that water is present to form oxide enough to saturate the acid; but, setting aside the inference I have drawn, that no water can be present with muriatic acid gas, the oxygen required to form the oxide in this case as well as the former, if derived from water, would evolve at least twice as much hydrogen. For, the relation of the oxygen in the oxide to the acid in the muriate, is proved by the fact, to be the same in the two cases.

Mr. Davy has, indeed, endeavoured to obviate any objection that may be made, as to the source of the oxygen in these experiments; he has found that nearly the same weight of muriate of mercury is formed, by precipitating a mercurial solution by a given volume of muriatic acid gas, as by burning potasium in the same quantity of gas, and then transferring the acid to mercury: he observes, 'there was no notable difference in the results.' The inference must, I conceive, be erroneous; 100 cubic inches of muriatic acid gas, united to potash, must give more muriate of potash, than if potasium was burned in the same gas; the weights of the materials necessarily require it; unless it be found that the two muriates are not the same salt.

From all the muriates, or salts, into which the muriatic acid enters, it appears (as will be shewn when these salts are considered) that the weight of an atom of muriatic acid is 22 times that of hydrogen. Very soon after this determination, it occurred to me that hydrogen was probably the base of the acid; if so, an atom of the acid must consist of 1 atom of hydrogen and 3 atoms of oxygen, as the weights of these just make up 22. In 1807 this idea was announced, and a suitable figurative representation of the atom was given, in the Chemical Lectures at Edinburgh and Glasgow; but this constitution of the acid was hypothetical, till these experiments of Mr. Davy seem to put it past doubt. The application of the theory to the experiments is as follows: on the supposition that the specific gravity of muriatic acid gas is 1.67, it will be found that 12 measures of the acid contain 11 measures of hydrogen, if liberated, and about 16½ measures of oxygen; then if ¼th of the acid be decomposed, nearly 3 measures of hydrogen will be liberated, and 4 + measures of oxygen, and the atoms of this oxygen will apply, 1 to 1, to the atoms of potasium, and furnish potash for the remaining ¾ths of the acid, (because 1 atom of acid contains 3 of oxygen). The very same explanation will apply to the formation of muriate of mercury. Here the hydrogen will be rather less than ¼th of the volume of the acid gas; but if we adopt Kirwan's specific gravity of muriatic acid, 1.93, then the hydrogen evolved will be between ⅓ and ¼th of the volume of acid gas.

Hence we conclude that an atom of muriatic acid gas consists of 1 atom of hydrogen and 3 of oxygen, or 1 atom of water and 2 of oxygen, and its weight = 22. Moreover, the diameter of the acid atom will be found (page 226) = 1.07, that of hydrogen being 1; or 12 measures of acid contain as many atoms as 11 measures of hydrogen, or as 5½ of oxygen.

My own experiments on muriatic acid gas have not been productive of important results. I sent 1000 small shocks of electricity through 30 measures of gas; there was a diminution of

1 measure, and on letting up water the whole was absorbed, except one measure, which appeared to be hydrogen. I sent 700 shocks through a mixture of muriatic acid gas and hydrogen; there was no change. A mixture of muriatic acid gas and sulphuretted hydrogen being electrified, hydrogen was evolved, and sulphur deposited, but no change of volume. It was evident the sulphuretted hydrogen only was decomposed. When a mixture of oxygen and hydrogen is fired along with muriatic acid gas, water is formed, and it instantly absorbs nearly its weight of acid gas. From these and such like unsuccessful attempts to decompose the muriatic acid, the importance of Mr. Davy's experiments is manifest.

The relation of muriatic acid to water must now be considered. It has been stated that water at the common temperature and pressure, absorbs 400 or more times its bulk of the acid gas; that is, rather less than its own weight. Now, 3 atoms of water weigh 24, and 1 atom of the acid gas weighs 22; it seems probable, then, that the strongest liquid acid that can well be exhibited, is a compound of 1 atom of acid and 3 of water, or contains about 48 per cent. acid. It is seldom sold of more than half this strength. Mr. Kirwan's table of the strength of muriatic acid of different specific gravities is very nearly correct; which, with some little addition and modification, is as follows:

Table of the quantity of real acid in 100 parts of liquid muriatic acid, at the temperature 60°.

Atoms.		Acid per cent. by weight.	Acid per cent. by measure.	Specific Gravity.	Boiling Point.
Acid.	Water.				
1 +	1	73.3	—	—	
1 +	2	57.9	—	—	
1 +	3	47.8	71.7 ?	1.500 ?	60°
1 +	4	40.7			
1 +	5	35.5			
1 +	6	31.4			
1 +	7	28.2			
1 +	8	25.6	30.5	1.199	120°?
1 +	9	23.4	27.5	1.181	145°?
1 +	10	21.6	25.2	1.166	170°
1 +	11	20.0	23.1	1.154	190°
1 +	12	18.7	21.4	1.144	212°
1 +	13	17.5	19:9	1.136	217°
1 +	14	16.4	18.5	1.127	222°
1 +	15	15.5	17.4	1.121	228°
1 +	20	12.1	13.2	1.094	232°
1 +	25	9.91	10.65	1.075	228°
1 +	30	8.40	8.93	1.064	225°
1 +	40	6.49	6.78	1 047	222°
1 +	50	5.21	5.39	1.035	219°
1 +	100	2.65	2.70	1.018	216°
1 +	200	1.36	1.37	1,009	214°

The first column shews the number of atoms of acid and water which are found combined in liquid acids of the different specific gravities; the second contains the acids per cent. by weight; that is, 100 grains of the liquid acid contain so many grains of pure acid ; the third contains the grains of acid in 100 water grain measures ; this is convenient in practice to prevent the trouble of weighing the acid ; the fourth contains the specific gravity of the liquid acid ; and the fifth contains the temperature at which acids of the various strengths boil. This last is entirely new, I apprehend ; it shews a remarkable gradation of temperature : the strong acid boils at a moderate heat ; as the acid weakens, the boiling temperature rises till it gets to 232° ; after which it gradually drops again to 212°. When an acid below 12 per cent. is boiled, it loses part of its quantity, but the remainder, I find, is concentrated ; on the other hand, an acid stronger than 12 per cent. is rendered more dilute by boiling. It appears from a paper of Dr. R. Percival in the 4th vol. of the Irish Transactions, that in the ordinary process of manufacturing the muriatic acid, the middle product is usually of the strength which boils at the maximum temperature ; but the first and last products are much stronger. The reasons for these facts will probably be found in the gradation of temperature in the above column.

5. *Oxymuriatic Acid.*

The highly interesting compound, now denominated oxymuriatic acid, was discovered by Scheele, in 1774. It may be procured by applying a moderate heat to a mixture of muriatic acid and oxide of manganese or red lead; a yellowish coloured gas ascends, which may be received over water; it is oxymuriatic acid gas. But this gas, which is largely obtained for the purposes of bleaching, is usually got from a mixture of equal weights of common salt (muriate of soda), oxide of manganese, and a dilute sulphuric acid of the strength 1.4; a heat at least equal to that of boiling water, seems required for the expulsion of the whole of the acid gas. Some of its properties are:

1. It has a pungent and suffocating smell, exceeding most other gases in these respects, and it is highly deleterious. Its specific gravity I find to be 2.34, that of common air being 1. Or, 100 cubic inches of it, at common pressure and temperature, weigh $72\frac{1}{2}$ grains.

2. Oxymuriatic acid gas is absorbed by water, but in a very small degree compared with muriatic acid gas. I find that at the

temperature of 60° and common pressure of pure gas, water takes up about twice its bulk of the gas. If the gas be diluted with air, then much less is absorbed, but the quantity is not proportionate to the abstract pressure of the gas, as is the case with those gases mentioned at page 201. Thus, if the pressure of oxymuriatic acid gas be $\frac{1}{7}$th of atmospheric pressure, water will be found to take up $\frac{2}{3}$ds of its bulk, which is more than twice the quantity it ought to take by the rule of proportion. Hence it is evident, that the absorption of this gas by water, is partly of a mechanical and partly of a chemical nature.

3. Water impregnated with the gas is called liquid oxymuriatic acid. It has the same odour as the gas, and an astringent, not acid, taste. When exposed to the light of the sun, the liquid acid is gradually decomposed, as was first observed by Berthollet, into its elements, muriatic acid and oxygenous gas; the former remains combined with the water, and the latter assumes the gaseous form. Neither light nor heat has been found to decompose the acid gas.

4. This acid, in the gaseous state or combined with water, has a singular effect on colouring matter. Instead of converting vegetable blue into red, as other acids do, it abstracts colours

in general from bodies, leaving them white or colourless. The oxygen combines with the colouring principle, and the muriatic acid remaining dissolves the compound. Hence the use of this acid in bleaching.

5. Combustible bodies burn in oxymuriatic acid gas more quickly than in common air, and the combustion is attended with several remarkable phenomena. Some bodies spontaneously take fire in this gas. All the metals are oxidized by this acid, and afterwards dissolved, forming salts denominated *muriates.* The combustible gases, mixed in due proportions with this acid gas, are either burned immediately, as sulphurous acid, sulphuretted hydrogen, nitrous gas, &c. or the mixture is capable of being exploded by an electric spark, as hydrogen, carburetted hydrogen, &c. These facts shew that the oxygen which combines with muriatic acid to form oxymuriatic, is easily abstracted again to enter into almost any other combination.

6. Oxymuriatic acid seems to combine readily with the fixed alkalis and the earths when dissolved in water; but it decomposes ammonia. It is remarkable, however, that few, if any, neutralized dry salts are to be obtained. When the saturated solutions are evaporated and crystallized, two distinct salts are chiefly

obtained; the one a simple muriate, and the other a hyperoxygenized muriate, in which an acid with an enormous quantity of oxygen is found, and is hence called *hyperoxymuriatic acid.*

7. One very remarkable property of oxymuriatic acid has recently occurred to me in a course of experiments upon it. Cruickshanks had found that if hydrogen and oxymuriatic acid gases were mixed together, and kept in a well stopped bottle for 24 hours, when the stopper was withdrawn under water, the gases disappeared, and water took their place. Being desirous to ascertain the time more definitely, I made the mixture in a narrow eudiometer, and left it to stand over water; in about three quarters of an hour the greater part of the mixture had disappeared. In the next experiment, the gases, after being put together, seemed to have no effect for one or two minutes, when suddenly the mixture began to diminish with rapidity, like one of common air and nitrous gas, except that there were no red fumes. The diminution went on, till in two or three minutes *nearly* the whole had disappeared. On repeating the experiment a few hours afterwards no such diminution was observed. I recollected that the sun had shone upon the instrument in the former one; it was

again placed in the direct rays of the sun, and the diminution was rapid as before. Upon repeating the experiment with sundry variations, it was confirmed, that *Light* is the occasion of this rapid combustion of hydrogen in oxymuriatic acid gas ; that the more powerful the light, the more rapid is the diminution of the mixture ; and that if the eudiometer be covered by an opake body, the mixture will scarcely be affected with any diminution for a day, and will not completely disappear in two or three weeks. Moreover, when the diminution is going on with speed, if the hand, or any opake body, is interposed to cut off the solar light, the diminution is instantly suspended. These observations equally apply to mixtures of carburetted hydrogen and carbonic oxide with the acid gas, except that the former deposits some charcoal. Carbonic acid, water, and muriatic acid, are the results.—These facts were ascertained in June 1809. In the ensuing month, I found that upon mixing hydrogen and oxymuriatic acid in a strong phial capable of containing 600 grains of water, and exposing the mixture to the solar rays, an explosion almost instantly took place with a loud report, just as if it had received an electric spark. If the stopper was well closed, a vacuum nearly was formed, which was instantly filled with water when the stopper was drawn out under water ; but it generally happened that the stopper was expelled with violence.

It remains now to point out the constitution of this acid. All experience shews, that it is a compound of muriatic acid and oxygen ; but the exact proportion has not hitherto been ascertained. Berthollet, who investigated the subject by impregnating water with the acid gas, and then exposing it to the solar rays till the oxygen was liberated, found it to consist of 89 parts of muriatic acid, and 11 of oxygen, by weight. Whether all the oxygen is liberated in this way is more than doubtful; the quantity of oxygen is certainly much underrated. Chenevix makes 84 muriatic acid and 16 oxygen to constitute this acid ; he too has the oxygen too low ; probably because he estimated *all* the salt formed by this acid to be simple muriate, or hyperoxymuriate ; but there is no doubt that oxymuriate does exist in the mixture, because it possesses the property of bleaching. Of all the authors I have seen, Cruickshank comes the nearest to the truth ; he says, 2 measures of hydrogen require 2.3 measures of oxymuriatic acid to saturate them ; and it is known that they require 1 of oxygen ; hence he infers, that 2.3 measures of this acid gas contain 1 measure of oxygen. From this it may be inferred, that 100 measures of the acid gas would afford 43.5 measures of oxygenous gas, and a certain unknown measure of muriatic acid (not 56.5, as Dr. Thomson has inferred). Chenevix remarks, that Cruickshank's gas was obtained from hyperoxymuriate of potash, and that 'the substance he ob-'tained was, in fact, not oxygenized muriatic 'acid gas, but a mixture of that gas with hy-'peroxygenized muriatic acid.' Dr. Thomson observes, that 'when water, impregnated with 'oxymuriatic acid gas, obtained by Cruick-'shank's method, is mixed with liquid ammo-'nia, scarcely any gas is extricated. The two 'bodies combine and form a salt.' I do not know what reasons these two authors may have had for making these remarks; but, according to my experience, they are entirely without foundation. The acid gas obtained from a mixture of sulphuric acid, muriate of soda, and manganese, or from muriatic acid and manganese, or from hyperoxymuriate of potash and muriatic acid, are all precisely the same, whether we consider their action upon the combustible gases, upon liquid or aeriform ammonia, or their absorbability by water. There is indeed one small difference, but it does not seem productive of any material effect; the gases obtained by the two former methods always deposit some brown oxide of manganese when treated with ammonia, but that obtained by the last deposits none. The action of muriatic acid on hyperoxymuriate of potash, evidently consists in detaching the superfluous oxygen from the compound, and not the hyperoxymuriatic acid particle from the particle of potash.

As the oxymuriatic acid is of great and increasing importance in a theoretical as well as practical point of view, I have spent much time in endeavouring to ascertain the proportion of its elements, and have, I think, succeeded; at least, I am pretty well satisfied myself as to its constitution: the methods I have taken are both synthetical and analytical, but I chiefly rely upon the latter.

1. I filled a eudiometer with dry mercury, and sent up 13 water grain measures of muriatic acid gas, to which were added 9 measures of oxygenous gas of 77 per cent. purity, which consequently consisted of 7 oxygen and 2 azote. The instrument had platina wires. About 1300 small electric shocks were passed through the mixture of gases; a gradual diminution ensued; the mercury became foul, the same as when oxymuriatic acid is in contact with it. The 22 measures were reduced to 4, which were not diminished by washing. To these 4

measures, 20 hydrogen and 20 common air were added, and the mixture being exploded, the diminution was 15 measures, corresponding to 5 oxygen; but the common air contained only 4 oxygen; therefore, 1 measure of oxygen must have been in the residuary gas, and probably 1 of azote was originally in the muriatic acid. Here then, it seems, 12 measures of muriatic acid united to 6 measures of oxygen to form oxymuriatic acid.—If we calculate from the specific gravities of the three elastic fluids, it will appear that 12 measures of muriatic acid gas, + 6 measures of oxygen gas, ought to make 11 measures of oxymuriatic acid gas. This result is nearly right; but the process is too laborious to be often repeated, especially as the object can be obtained much more easily and elegantly by the analytic method.

2. Oxymuriatic acid gas and hydrogen, mixed together over water, explode with an electric spark, much like a mixture of common air and hydrogen. Cruickshank mixed 3 measures of hydrogen with 4 of the acid, and exploded them over mercury: in this case, there was a residuum of acid gas. He then mixed 4 measures with 4, and after the explosion found a residuum of hydrogen. From these experiments, he infers, that 3 measures

of hydrogen require $3\frac{1}{2}$ of the acid to saturate them. I have found the results a little different; but the error is not much, and is what might be expected. Whether we treat oxymuriatic acid over mercury or water, we are sure to lose some of it; and unless the loss can be estimated and allowed for, we are apt to overrate the acid required. Before the action of light on this mixture was discovered, I used to mix known quantities of the two gases together, in a graduated eudiometer of Volta, over water; and, after letting the mixture stand a few minutes, in order to a complete diffusion, I passed a spark through, but noticed the moment before at what degree the mixture stood; in this way, when there is an excess of hydrogen, the results are accurate; the total diminution can be found, and the residuary gas can be analyzed to find the hydrogen left, and the common air (if any), which is extremely apt to be found in a greater or less degree, in all oxymuriatic acid obtained over water. By frequent careful trials, I found that a measure of hydrogen required as near as possible an equal measure of the acid to saturate it. But since the effect of solar light was discovered, I have operated in a more simple and elegant manner; and the results appear rather more uniform and accurate. I take a graduated tube, capable of containing 200 measures of gas. I fill this with water, and transfer into it 100 measures of hydrogen of known purity; to this a quantity of acid gas is added, so as to fill the tube nearly. The finger is then applied to the end of the tube, and it is instantly transferred to a jar of mercury. The whole is then taken, and exposed to the sun, (if not shining too powerfully, in which case an explosion may be apprehended) or to the strongest light that can be obtained; when, after remaining two or three minutes without exhibiting any change, the water, and afterwards the mercury, ascend the tube with increasing and then diminishing velocity, till they nearly reach the top. The residuary gas may then be examined, and the quantity of hydrogen, acid and common air, ascertained. The quantity of water in the tube becomes visible as the mercury ascends, and is useful to prevent the action of the acid on the mercury. The water must be subtracted from the capacity of the tube, to find the volume of gases employed, from which taking the hydrogen, there remains the acid, &c.

From the mean of five experiments executed as above, I am induced to conclude, that 100 measures of hydrogen require 94 measures of oxymuriatic acid gas to convert them into water. In every one of the experiments, the acid was less than the hydrogen.

The above experiments are highly amusing in a day of clouds and gleams; the presence of the direct solar light instantly gives the motion of the mercury a stimulus, and it as quickly abates when a cloud intervenes. The surface of the mercury in the tube always becomes fine sky blue during the process; and so does liquid ammonia that has been used to decompose oxymuriatic acid; I do not know what is the reason in either case.

From the results above, it appears that 100 measures of oxymuriatic acid gas must consist of 53 measures of oxygen, united to a certain portion of muriatic acid gas. Now, 100 cubic inches of oxymuriatic acid gas weigh 72 or 73 grains, and 53 inches of oxygen weigh about 18 grains, which is rather less than $\frac{1}{4}$th of the above. Hence, if the atom of muriatic acid weigh 22, that of oxymuriatic acid must weigh 29; and thus we obtain the constitution of this last acid. An atom of it consists of one of muriatic acid and one of oxygen united; the former weighs 22, the latter 7, together making 29; or about 76 muriatic acid, and 24 oxygen, per cent. Thus, it appears, that the former experiments on the specific gravities of those fluids, corroborate the recent ones on their constitution. If the constitution of muriatic acid be rightly determined, then oxymuriatic acid must consist of 1 atom of hydrogen and 4 of oxygen. At all events, 1 atom of muriatic acid must combine with 1 of oxygen to form 1 of oxymuriatic acid. The diameter of the elastic atom of this gas is nearly the same as hydrogen, and may therefore be denoted by 1, but it is rather less; and the number of atoms in a given volume of this gas is to the number in the same volume of hydrogen, as 106 to 100 nearly. It appears, then, that the atoms of oxymuriatic acid are rather more dense than those of muriatic acid, or than those of hydrogen.

5. *Hyperoxymuriatic Acid.*

The existence of a compound denominated hyperoxymuriatic acid, has been clearly shewn in a state of combination; but it has not, and perhaps can not, be exhibited in a separate, elastic, or even liquid form, probably on account of the great weight and number of its elementary parts. It is clearly a compound of muriatic acid and an enormous quantity of oxygen. It is obtained in combination with the alkalies and earths, by sending a stream of oxymuriatic acid gas into solutions of these elements, or of their carbonates in water. The acid combines with the alkali; but in process of time, as the solution becomes concentrated, a change takes place in the acid; one atom of oxymuriatic acid seizes upon an atom of oxygen from each of its neighbouring particles, and reduces them to ordinary muriatic acid; in this state it forms with an atom of alkali an hyperoxymuriate, whilst the other atoms of acid form muriates. It seems that the oxymuriates are difficultly attainable; because, as their solutions are concentrated, they are so apt to be resolved and compounded again, as above.

Berthollet first pointed out the peculiarity of this acid: but its nature and properties were more fully discussed by Hoyle in 1797, and by Chenevix in 1802. These authors made their principal experiments on hyperoxymuriate of potash; they nearly agree as to the constitution of the salt, but differ in some of the circumstances of its production. It yields by heat about 2 or 3 per cent. of water, about 38 per cent. of oxygen, and 59 or 60 of a salt unalterable by heat, which Chenevix considers as simple muriate; but Hoyle says it exhibits traces of oxymuriatic acid by sulphuric acid.

The acid in 59 muriate is nearly 20. Hence, 20 muriatic acid added to 38 oxygen by weight, constitute 58 of hyperoxymuriatic acid: or, as Chenevix states it, 65 oxygen + 35 muriatic acid = 100 hyperoxymuriatic acid. This I judge to be very nearly true. Now, if 35 muriatic acid require 65 oxygen, 22 will take 41; but 22 is the weight of an atom of muriatic acid, and 41 or 42 is the weight of 6 atoms of oxygen; hence the constitution of hyperoxymuriatic acid is determined. An atom of it consists of 1 atom of muriatic acid + 6 atoms of oxygen, or of 1 atom of oxymuriatic acid + 5 atoms of oxygen; and its weight is represented by 64. We may now see what takes place in the formation of hyperoxymuriates. One atom of oxymuriatic acid deprives 5 surrounding atoms, each of an atom of oxygen; an atom of hyperoxymuriate thus necessarily produces 5 atoms of simple muriate. Supposing the salts from potash, their weights may be found thus: An atom of potash weighs 42, one of hyperoxymuriatic acid weighs 64, together = 106. Five atoms of muriate of potash = 320; the sum of both = 426. Now, if 426 : 106 : : 100 : 25 nearly. Hence, in the formation of hyperoxymuriate of potash, if the whole potash is formed into muriate and hyperoxymuriate, there must be 75 of the former and 25 of the latter. Hoyle does not inform us on this head; Chenevix found 84 of the former and 16 of the latter. Here then is some obscurity. The fact, I believe, is, that there is always a greater or less portion of real oxymuriate of potash amongst the salts formed, or in the mass which Chenevix calls the *entire salt*. Oxymuriatic acid precipitates silver from nitrate as well as muriatic; and as this was the test, it is evident Chenevix must have confounded a quantity of oxymuriate of potash with the muriate. The quantity may even be ascertained. For, if 25 : 75 :: 16 : 48. In 100 of Chenevix's entire salt, there were then 16 hyperoxymuriate, 48 muriate, and the rest or 36 must have been oxymuriate. Hoyle's experiments confirm this conclusion; for, he observes that the remaining muriate (after the hyperoxymuriate was abstracted) was considerably oxygenized, since with the addition of acids it became a powerful destroyer of vegetable colours. This could not be the case with a muriate, nor even a mixture of muriate and hyperoxymuriate. Besides, it is well known that the oxymuriate of potash (or oxymuriatic acid absorbed by potash) was largely used for the purpose of bleaching; now if the acid had immediately resolved itself into muriatic and hyperoxymuriatic, it would have been of no use for that purpose.

Hyperoxmuriatic acid must then be constituted of 1 atom of muriatic acid and 6 of oxygen; but as the former is probably composed of 1 atom of hydrogen and 3 of oxygen, we have 1 atom of hydrogen + 9 of oxygen for the constitution of an atom of the first mentioned acid; or it consists of $1\frac{1}{2}$ hydrogen + $98\frac{1}{2}$ of oxygen per cent. by weight. It is no wonder, then, if this acid readily part with its oxygen, and be apt to explode when treated with combustible bodies; nor if it refuse to form an elastic fluid of such unwieldy particles.

Note on Fluoric and Muriatic Acids.

Since the foregoing articles on fluoric and muriatic acid were printed off, I have seen the Journal de Physique, for January 1809, in which is an abstract of an highly interesting Memoir on the Fluoric and Muriatic Acids, by Gay-Lussac and Thenard. Their observations, supported by facts, are remarkably in unison with those I have suggested. They find that when fluoric acid gas is admitted to any gas, and produces fumes, the gas is diminished, but only a small quantity; that when no fumes appear, no diminution takes place; they hence conclude, that this acid gas is an excellent test of the presence of hygrometric water [steam] in gases; and observe that all gases contain such, except fluoric, muriatic, and probably ammoniacal. Berthollet, jun. has proved the last mentioned gas to contain no combined water; and Gay-Lussac and Thenard suspect it contains none hygrometrically; but some experiments of Dr. Henry convince me that it does; and I think its not fuming when mixed with common air is a proof of it.—They observe, that when water is saturated with fluoric acid gas, it is limpid, smoking, and extremely caustic; that heat expels about one fifth of the acid, and the remainder becomes fixt, resembling concentrated sulphuric acid, and requiring a high temperature to boil it. They query from this fact, whether sulphuric and nitric acid are not naturally gasiform, and owe their liquidity to the water combined with them. They exposed a drop of water to 60 cubic inches of fluoric acid gas; the drop, instead of evaporating, was increased in volume by the absorption of the acid; and hence they conclude, that fluoric acid gas is also free from combined water; the conclusion is extended to ammoniacal gas, but not to muriatic acid gas. I wonder at their exception with regard to muriatic acid, as every one knows it presents the same phenomena when a drop of water is admitted; that is, the drop is increased by the condensed acid, and suffers no evaporation. They allude, however, to the experiments of Henry and Berthollet, in which water was supposed to be found in a state of intimate union with this acid gas; and they mention some of their own, in which one fourth of the weight of the gas was found to be water. This conclusion of muriatic acid gas being the only gas that contains water combined with it, they consider as striking; and seem inclined to consider water as a constituent of the acid, but that the oxygen and hydrogen are not in the state of water.

Gay-Lussac and Thenard found that fluoric acid gas, detached from fluate of lime by boracic acid, does not dissolve silica, on account of the boracic acid which it holds in solution. Another remarkable fact was, that fluate of lime, distilled with sulphuric acid in leaden vessels, does not give the fluoric acid in an elastic, but in a liquid form.—They observe, as Davy had done, that in burning potasium in siliceous fluoric acid gas, some hydrogen is given out, amounting successively to about one third of what would be given out by water.

They seem to think that the acid is decomposed in this case: but they have not advanced any opinion, that either fluoric or muriatic acid gas consists entirely of hydrogen and oxygen.

SECTION 2.

OXYGEN WITH AZOTE.

The compounds of oxygen with azote, hitherto discovered, are five; they may be distinguished by the following names; nitrous gas, nitric acid, nitrous oxide, nitrous acid, and oxynitric acid. In treating of these, it has been usual to begin with that which contains the least oxygen, (nitrous oxide) and to take the others in order as they contain more oxygen. Our plan requires a different principle of arrangement; namely, to begin with that which is most simple, or which consists of the smallest number of elementary particles, which is commonly a binary compound, and then to proceed to the ternary and other higher compounds. According to this principle, it becomes necessary to ascertain, if possible, whether any of the above, and which of them, is a binary compound. As far as the specific gravities of the two simple gases are indicative of the weights of their atoms, we should conclude that an atom of azote is to one of oxygen as 6 to 7 nearly; the relative weights of ammonia and water also give countenance to such a ratio. But the best criterion is derived from a comparison of the specific gravities of the compound gases themselves. Nitrous gas has the least specific gravity of any of them; this indicates it to be a binary compound; nitrous oxide and nitrous acid are both much heavier; this indicates them to be ternary compounds; and the latter being heavier than the former, indicates that oxygen is heavier than azote, as oxygen is known to abound most in the latter. Let us now see how far the facts already known will corroborate these observations.

According to Cavendish and Davy, who are the best authorities we yet have in regard to these compounds, they are constituted as under:

	Sp. gr.	constitution by weight.	Ratios.	
Nitrous gas	1.102	46.6 azote + 53.4 oxy.	6.1:7	Davy.
		44.2 —— + 55.8 ——	5.5:7	
		42.3 —— + 57.7 ——	5.1:7	
Nitr. oxide	1.614	63.5 —— + 36.5 ——	2×6.1:7	
		62 —— + 38 ——	2×5.7:7	
		61 —— + 39 ——	2×5.4:7	
Nitric acid	2.444	29.5 —— + 70.5 ——	5.8:7×2	
		29.6 —— + 70.4 ——	5.9:7×2	Cavendish.
		28 —— + 72 ——	5.4:7×2	
		25.3 —— + 74.6 ——	4.7:7×2	

The above table is principally taken from Davy's Researches: where two or more results are given under one article, they are derived from different modes of analysis. In the third column are given the ratios of the weights of azote and oxygen in each compound, derived from the preceding column, and reduced to the determined weight of an atom of oxygen, 7. This table corroborates the theoretic views above stated most remarkably. The weight of an atom of azote appears to be between 5.4 and 6.1: and it is worthy of notice, that the theory does not differ more from the experiments than they differ from one another; or, in other words, the mean weight of an atom of azote derived from the above experiments would equally accommodate the theory and the experiments. The mean is 5.6, to which all the others might be reduced. We should then have an atom of nitrous gas to weigh 12.6, consisting of 1 atom of azote and 1 of oxygen; an atom of nitrous oxide to weigh 18.2, consisting of 2 atoms of azote and 1 of oxygen; and an atom of nitrous acid to weigh 19.6, consisting of 1 atom of azote and 2 of oxygen. Nor has the weight of an atom of oxygen any influence on the theory of these compounds; for, it is obvious that if oxygen were taken 3, or 10, or any other number, still the ratios of azote to oxygen in the compounds would continue the same; the only difference would be, that the weight of an atom of azote would rise or fall in proportion as that of oxygen.

I have been solicitous to exhibit this view of the compounds of azote and oxygen, as derived from the experience of others, rather than from my own; because, not having had any views at all similar to mine, the authors could not have favoured them by deducing the above results, if they had not been conformable to actual observation.

I come now to make some observations on the results contained in the preceding tables, and to state those of my own, which have been obtained with labour and assiduity.

I believe the above mean weight of an atom of azote, 5.6, is too large; and that the true mean is but little above 5; perhaps 5.1, or 5.2.—I do not mean by this observation to insinuate that the results in the above table are derived from inaccurate experiments. In the course of my investigations, I have had to repeat the experiments of many; but have found no results to which my own in general approximated so nearly as to those of Mr. Davy in his Researches. As knowledge advances, however, greater precision is attainable from the same facts. As for Mr. Cavendish's important experiments, they were intended to shew what elements constitute nitric acid, rather than the proportion of them; and they were made at too early a period of pneumatic chemistry to obtain precision.

The first line of the table contains the proportions of azote and oxygen in nitrous gas, as determined by the combustion of pyrophorus. Mr. Davy justly considers this as least entitled to confidence. The second and third were obtained from the combustion of charcoal in nitrous gas. The second is grounded upon the oxygen found in the carbonic acid. By making the calculation of this from more recently determined proportions of charcoal and oxygen, I reduce the azote to 5.4. The third is derived from the azote left after combustion. Mr. Davy finds 15.4 measures of nitrous gas yield 7.4 of azote; or 100 measures of nitrous gas yield 48 measures of azotic gas.

Dr. Priestley was the first to observe that the electric spark diminishes nitrous gas, and finally leaves azotic gas; he states the reduction to be to one fourth of the volume. I have several times repeated this experiment with all possible attention to accuracy; the exact quantity of azote in the nitrous gas was previously determined by sulphate of iron, and was commonly 2 per cent.; the quantity of 50 or 100 water grain measures of the gas was put into a narrow eudiometer tube over water, furnished with platina wires; the electrification was for one or two hours, and uninterruptedly continued till no further diminution was observable. To the residuary gas a small portion of common air was added, and no diminution found. In this way, from 100 measures of pure nitrous gas there are obtained at a mean 24 measures of azotic gas; or, which is the same thing, 102 measures of the 98 per cent. gas leave a residuum of 26 azote. The deviation was never more than 1 per cent. from the above; that is, from 100 measures of pure nitrous gas I never obtained more than 25 measures of azote, nor less than 23. I believe, therefore, that 24 measures may be safely relied upon as an accurate approximation.

This experiment, taken in conjunction with the last mentioned one of Mr. Davy, is of great importance. It not only shews the constitution of nitrous gas, but that of nitric acid also. It appears, that by electrification exactly *one half* of the azotic gas is liberated; and its oxygen joins to the *other half* to form nitric acid. The immediate effect of the electric shock is to separate the atoms of azote and oxygen, which by their junction form nitrous gas; the moment the oxygen is liberated, it is seized by another atom of nitrous gas, and the two united form an atom of nitric acid which escapes into the water. In other words, 100 measures of nitrous gas conrain 48 of azote; by electrification, 24 measures of azote are liberated, and the other 24 measures acquire the oxygen lost by the former, and become nitric acid, which are absorbed by the water.

A repetition of Mr. Cavendish's experiments will be found to confirm the above conclusion. I have in three or four instances undertaken experiments of the same nature, and with like results; but as these are of a laborious kind, it is not so convenient to execute them. One of these was more particularly an object of attention, and I shall relate it in the detail. A quantity of pure oxygenous gas was diluted with common air by degrees till the mixture contained 29 measures per cent. of azote, that being presumed to be nearly the due proportion to form nitric acid. The test was, exploding it with hydrogen, and taking $\frac{1}{3}$ of the diminution for oxygen. A portion of distilled water was impregnated with this mixture of gases, and put into a eudiometer furnished with platina wires. Into this, 50 measures of the mixed gases were put, and the electrification commenced; after several hours electrification, it was reduced to 20 measures; it continued there all night without any change, the operation was resumed next day, and the gas was reduced to 13 measures. These were found to be $3\frac{1}{2}$ azote + $9\frac{1}{2}$ oxygen; or 27 azote + 73 oxygen per cent. Hence it was evident, that 29 measures per cent of azote were too small; by calculation from the above data, it will be found that 30 measures of azote unite to 70 of oxygen to form nitric acid. This gives 27 of azote by weight, and 73 of oxygen in nitric acid; which nearly agrees with the mean of Cavendish. From this, the weight of an atom of azote comes out 5.15 —By the experiment on nitrous gas, supposing its specific gravity 1.10, and that of azote 966, the weight of an atom of azote comes out 5.1.

With respect to nitrous oxide, I think Mr. Davy's calculations scarcely do justice to his experiments. The first line shews the results derived from the combustion of hydrogen in nitrous oxide. From several experiments, Mr. Davy selects one in which 39 measures of nitrous oxide and 40 of hydrogen were fired together, and seemed just to saturate each other, leaving a residuum of 41 azote; but this residuum must have contained a few atoms of azote originally mixed with the oxide and the hydrogen, and may therefore be supposed to be overrated. If we suppose 39 oxide to contain 40 azote, it will reduce the weight of an atom of azote from 6.1 to 5.6. In my own experience, equal volumes of nitrous oxide and hydrogen, saturate each other, and the volume of azote left is equal to one of the other two, making the due allowance for impurities. This would imply that a measure of azote + half a measure of oxygen, should, when combined, constitute a measure of nitrous oxide; but the united weights are about 5 per cent. too little, according to the specific gravity of the oxide given above. I apprehend the oxygen this way is underrated, owing perhaps to the formation of an unperceived quantity of nitric acid. In the second line, we have the proportions of azote and oxygen in nitrous oxide, derived from the combustion of both phosphuretted hydrogen and charcoal in the oxide. By the former, nitrous oxide gave an equal volume of azote; by the latter, 21 measures of oxide produced 21.5 measures of azote, and 11.5 measures of carbonic acid. Now, if we suppose that a measure of nitrous oxide contains an equal volume of azotic gas weighing .966, and the rest of the weight, .648 to be oxygen, the proportion will be 60 azote + 40 oxygen per cent. by weight. Further, it is now known that 11.5 measures of carbonic acid contain 11.5 measures of oxygen; hence 21 measures of nitrous oxide must contain 11.5 measures of oxygen; say 20 measures of oxide, because 30 being used in all, and 9 pure being abstracted from the residuum, the remainder 21 must have contained the impurities in all the 30 measures, which could scarcely be less than 1. This gives as before, 60 azote + 40 oxygen by weight per cent. in nitrous oxide. The third line gives the results obtained from the combustion of sulphuretted hydrogen; here Mr. Davy found 35 measures of nitrous oxide saturate 20 measures of sulphuretted hydrogen, and leave a residuum of $35\frac{1}{2}$ measures of azote: This seems again to shew that the azote is equal in volume to the oxide, and consequently will give as before, 60 azote + 40 oxygen, by weight; and the weight of an atom of azote will be accordingly found = 5.25.

It is remarkable, that in the combustion of hydrogen in nitrous oxide, the oxygen (as estimated by the loss of hydrogen) is usually found below par; and it is the same with the azote in the combustion of olefiant gas, as Mr. Davy has remarked; I have found it so likewise with carburetted hydrogen or coal gas. I apprehend when azote disappears, it is from the formation of ammonia.

Besides the three compounds of azote and oxygen already considered, there are at least two more. One is called *nitrous* acid; it is a compound of nitric acid and nitrous gas. The other I call oxynitric acid; it is a compound of nitric acid and oxygen. Priestley discovered the fact that nitric acid absorbs nitrous gas very largely, and thereby becomes more volatile. He found that 130 ounce measures of nitrous gas over water disappeared in a day or two, when a phial containing 96 water grain measures of strong nitric acid was inclosed with the gas. The colour of the acid as it absorbs nitrous gas is gradually changed from pale yellow to orange, green, and finally blue green. Mr. Davy has used his endeavours to find the quantity of nitrous gas which nitric acid absorbs; he estimates the blue green acid of 1.475 sp. gr. to contain 84.6 nitric acid, 7.4 water, and 8 nitrous gas, by weight; and he concludes that dilute acids absorb less nitrous gas in proportion than concentrated acids. This subject shall be presently considered.

Priestley discovered that nitrous gas entered into combination with oxygen upon the mixture of the two gases. In this way it is easy to saturate one of the gases with the other; but it unfortunately happens that two or three distinct compounds are usually formed, and the proportion of one compound to another varies according to the circumstances of the mixture. By the constitution of nitric acid above determined, it follows that 10 measures of oxygen will require 18 measures of nitrous gas to convert them into nitric acid. But the mixture may be so managed as that 10 of oxygen shall take either 13 or 36 measures, or any intermediate number. As the facts relating to this matter have not been distinctly stated by any author I have seen, I shall subjoin the results of my own experience.

1. When 2 measures of nitrous gas are put to 1 measure of oxygen, in a tube one third of an inch in diameter, and 5 inches in length, and as soon as the diminution is apparently

ceased, which will be half a minute, the residuary gas is transferred into another tube, it will be found that 1 measure of oxygen and 1.8 of nitrous gas have disappeared ; the mixture is to be made over water.

2. When 4 measures of oxygen are put to 1.3 of nitrous gas in a tube two tenths of an inch in diameter, and 10 inches long, so as to fill it ; it will be found that 1 measure of oxygen will combine with 1.3 of nitrous gas, in 4 or 5 minutes.

3. When 1 measure of oxygen and 5 of nitrous gas are mixed together, so as to form a thin stratum of air, not more than $\frac{1}{8}$th of an inch in depth (as in a common tumbler) ; it will be found that the oxygen will take from 3 to $3\frac{1}{2}$ measures of nitrous gas in a moment, and without any agitation. If equal measures are mixed, then 1 oxygen takes about 2.2 nitrous.

4. When water has been made to imbibe a given portion of oxygenous gas, and is afterwards agitated in nitrous gas, the quantity of nitrous gas absorbed will always be more than exhausted water would take, by a quantity equal to 3.4 or 3.6 times the bulk of the oxygenous gas. And, *vice versa*, when water has imbibed a portion of nitrous gas, and is then agitated with oxygenous gas, the quantity

absorbed will be greater than exhausted water would take, by a portion which bears to the nitrous the ratio of 1 to 3.6.

These facts are of a nature easily to be ascertained, and I have no doubt will be found near approximations to the truth, by those who may repeat them. They are curious and singular; as we have few other examples where two gases form a real chemical union in such varied proportions. If the gases be not mixed precisely as above in all the circumstances, the results will not be the same. But in all the variations I have observed, I have not found oxygen to be saturated with less than 1.3, nor with more than 3.6 measures of nitrous gas. It is obvious that the presence of water, and the shortness of the column of the mixed gases, both contribute to the great expenditure of nitrous gas; the latter probably from its suffering the union to take place instantaneously. On the other hand, a narrow tube makes the operation more slow, and removes the point of union far from the surface of the water; these circumstances seem to increase the quantity of oxygen entering into combination.

What then are we to conceive of this compound of oxygen and azote, in which 1 measure of oxygen sometimes combines with 1.3 of nitrous gas, and sometimes with 3.6, and according to circumstances takes any intermediate portion? Are there indefinite gradations in the compound? I cannot conceive this; neither do the facts at all require it. All the products that need be admitted to explain the facts are three. It has been shewn that 1 measure of oxygen requires 1.8 of nitrous gas to form nitric acid, according to the results derived from the electrification of nitrous gas; and the conclusion is corroborated by other facts. It appears from the above observations, 3 and 4, that oxygen is found sometimes to combine with 3.6 times its bulk of nitrous gas, and that this is the maximum; but it is just twice the quantity requisite to form nitric acid; it is evident, therefore, that a compound is formed in which there are twice as many atoms of nitrous gas as are necessary to form nitric acid. This then may be called *nitrous* acid; and the elementary atoms consist of 1 of oxygen and 2 of nitrous gas, united by chemical affinity. If the other extreme, or the minimum quantity of nitrous gas to which oxygen had united, had been .9, or half what is found in nitric acid, then this would have shewn the union of 2 atoms of oxygen with 1 of nitrous gas, and the compound might be called *oxynitric* acid. Now, though it does not appear that we are able as yet to form this compound exclusively, yet it is highly probable that it exists, and that it is always formed along with nitric acid, and perhaps even with nitrous acid, when the oxygen consumed is more than 1 measure for 1.8 of nitrous gas. When 1 measure of oxygen unites with 1.8 of nitrous gas, as mentioned in the first observation, I conceive it is not purely nitric acid that is formed, but a mixture of all the three acids, in such proportions that the nitrous and oxynitric balance each other, and in the sequel, when combined with water, these two become, by their interchange of principles, nitric acid.

We shall now proceed to remark more particularly on the different compounds of azote and oxygen: but it may not be amiss to state here in a table their constitution, as far as appears from the preceding views and observations.

	Wt. of an atom	Atoms of azote. ox.	100 parts by wt. contain azote. oxy.	100 parts by meas. contain azote. oxyg.
Nitrous gas	12.1 =	1+1	42.1 +57.9	48 + 56.6
Nitrous oxide	17.2 =	2+1	59.3 +40.7	99.1 + 58 3
Nitric acid	19.1 =	1+2	26.7 +73.3	30 : 70*
Oxynitric acid	26.1 =	1+3	19.5 +80.5	22.1 : 77.9
Nitrous acid	31.2 =	2+3	32.7 +67.3	36.2 : 63.8

* The specific gravities of the three last not being accurately determined, we can only give the *ratios* of the measures, and not the absolute quantities of azote and oxygen in 100 measures.

1. *Nitrous Gas.*

Nitrous gas is formed by pouring dilute nitric acid upon many of the metals; it should be received over water. The best mode of procuring it is to put a few small pieces or filings of copper into a gas bottle, and pour nitric acid of the specific gravity 1.2 or 1.3 on to them; the gas comes over in a state of purity (except so far as it is diluted with atmospheric air) and without the application of heat. The common explanation of this process is, that a part of the nitric acid is decomposed into the elements nitrous gas and oxygen; its oxygen unites to the metal to form an oxide, which the rest of the acid dissolves. Upon a more particular examination of the phenomena, I find, that estimating the quantity of real acid by Kirwan's table, $\frac{1}{3}$ part of the acid is decomposed to furnish oxygen to the metal, and to yield nitrous gas, $\frac{1}{3}$ unites to the metallic oxide, and the remaining $\frac{1}{3}$ seizes the nitrous gas, and forms nitrous acid; but in the degree of condensation of the acid, it is unable to hold more than $\frac{1}{3}$ or $\frac{1}{2}$ of it, and the rest is therefore evolved. For example, 200 grain measures of nitric acid of 1.32 strength, diluted with 100 water, dissolved 50 grains of copper, and yielded 44 cubic inches of nitrous gas = 15 grains. Now, 200 measures of the acid contained 102 grains of real acid; and 50 of copper require 35 of nitric acid, which is nearly $\frac{1}{3}$ of 102; every atom of copper takes two atoms of oxygen to form the oxide, and this oxide takes two atoms of nitric acid to form the nitrate of copper (as will be shewn in the sequel); whence it appears that whatever quantity of acid is employed to oxidize the copper, an equal quantity is required to unite to the oxide; the quantity of nitrous gas given out should therefore have been 22 grains, but it was only 15: it seems, then, that 7 grains of nitrous gas combined with the remaining acid to form *nitrous* acid, part of which was probably volatilized by the heat excited in the mixture.

Nitrous gas, according to Kirwan, has the specific gravity 1.19; according to Davy 1.102; this last is the nearest approximation to truth, as far as my experience goes. Its ultimate particle weighs nearly 12.1 of hydrogen; the diameter of it in an elastic state is .958, that of hydrogen being 1; if a measure of hydrogen contain 1000 atoms, the same measure of nitrous gas will contain 1136 atoms. This gas is highly deleterious when inspired in a dilute state; if pure, it is instantly fatal. It extinguishes combustion in general; but pyrophorus spontaneously takes fire in it; and phosphorus and charcoal in an ignited state burn in it, and produce a decomposition. Pure water, (that is, water free from all air) I find, absorbs about $\frac{1}{18}$th of its bulk of nitrous gas; but only $\frac{1}{27}$th of it can be expelled again by other gases: it should seem, then, that a small portion of the gas actually combines with the water, while the greater part is, like most other gases, mechanically retained by external pressure.

Nitrous gas, as has been observed, is decomposed by electricity: one half of the azote is liberated, and the other half unites with the evolved oxygen, and forms nitric acid. According to Davy's analysis by charcoal, nitrous gas is constituted of 2.2 azote, and 3 oxygen by weight; or 42 azote, and 58 oxygen per cent. nearly; which is the same as I obtain by electricity and other means. If completely decomposed, 100 measures would be expanded to 104.6, of which 48 would be azote, and 56.6 oxygen.

Dr. Henry has recently discovered that nitrous gas is decomposed by ammoniacal gas; the two gases are mixed over mercury in Volta's eudiometer, and an electric spark is found sufficient to explode them. When an excess of nitrous gas is used, the products are, azotic gas and water with a small portion of nitric acid ; when an excess of ammonia is used, then azotic gas, water, and hydrogen are produced. When ammoniacal gas is sent through a tube, containing manganese red hot, Dr. Milner found that nitrous gas was formed. These facts exhibit remarkable instances of the decomposition and composition of nitrous gas.

The degree of purity of nitrous gas is easily and accurately ascertained, by means of a strong solution of certain salts of iron, particularly the common sulphate or green copperas. A measure of the gas is put into a narrow tube, and the end of it dipped in the solution ; as soon as a small portion of the liquid has entered the tube, a finger is applied to the end, and the liquid is agitated ; the tube is again immersed in the liquid, and the finger withdrawn, when a portion of the liquid enters: the process is repeated till no more gas is absorbed. What remains is usually azotic gas. The absorption is rapid, and the operation completed in a minute. This fact was first observed by Dr. Priestley. Wishing to know the nature of this combination more minutely, I procured a solution of green sulphate, such that 6 grain measures contained

1 grain of the salt ; its specific gravity was 1 081 ; this was agitated with iron filings, to reduce any of the red sulphate that might be in the solution, which is known not to absorb the gas, into green sulphate. A eudiometer was filled with mercury, except one measure, which was filled with the liquid solution ; the tube was then inverted over mercury, and nitrous gas sent up to the solution, which was afterwards agitated. It was repeatedly found that 1 measure of the solution absorbed 6 measures of the gas, and was then saturated. Consequently 1500 grain measures of the solution would have taken 9000 grain measures of the gas ; but 1500 of the solution contained 250 of salt, of which $\frac{1}{5}$th was iron, as is well known ; and 9000 grain measures of the gas weigh 12 grains : Here, then, 50 grains of iron united to 12 grains of nitrous gas. Now, the weight of an atom of iron is 50 (page 258), and that of nitrous gas is 12. It therefore follows, that in the combination of green sulphate of iron with nitrous gas, each atom of iron unites with an atom of the gas, agreeably to the general law of chemical union.

Nitrous gas is still used in eudiometry to determine the quantity of oxygenous gas in any mixture ; and on account of the ease and elegance of its application, and the quickness with which it attaches that gas, it will always be used. It has been found, however, that the simple mixture of the two gases is not enough to discover the proportion of oxygen, by reason of the different compounds that are formed. The object may be effectually obtained, by using an excess of nitrous gas of a known strength, and then abstracting the surplus by means of sulphate of iron. Some authors prefer a solution of green sulphate of iron saturated with nitrous gas; the oxygenous gas is agitated in a portion of the solution, and the residuary gas is washed with a solution of the sulphate, unimpregnated with nitrous gas. But the quantity of oxygen in certain mixtures is ascertained with equal or greater precision, by firing it with hydrogen in Volta's eudiometer, and taking $\frac{1}{3}$ of the diminution for oxygen; or by agitating the gas in a small portion of sulphuret of lime, which abstracts the oxygen.

When nitrous gas is mixed with oxymuriatic acid gas over water, an instantaneous diminution of volume takes place. I was in expectation that this would convert the nitrous gas into pure nitric acid, and consequently the quantity of oxygen necessary would be ascertainable this way; but the two gases, like oxygen and nitrous gas, combine in various proportions, according as one or other is in excess. Sometimes 3 measures of nitrous are saturated with 2 of the acid, and sometimes with 4 measures. When green sulphate of iron is saturated with a known portion of nitrous gas, and the solution is afterwards agitated with oxygen, the absorption is somewhat slow, (like that with sulphuret of lime) and the quantity taken up is equal in bulk to the nitrous gas. The liquid, from a dark red or black, becomes of a bright yellowish red, the oxide of iron being changed from the green to the red during the process.

It has been made appear, that by electricity *one half* of the atoms of nitrous gas are decomposed, in order to oxygenize the other half; in like manner, in certain cases, *one half* of the atoms of nitrous gas are decomposed to *azotize* the other half. This is shewn by the experiments of Priestley, but much more accurately by those of Davy. The alkaline sulphites, muriate of tin, and dry sulphures, convert nitrous gas into nitrous oxide. According to Davy, 16 cubic inches of nitrous gas were converted into 7.8 of nitrous oxide by sulphite of potash ; that is, 100 measures gave 48.75: he also found, that muriate of tin and dry sulphures changed 100 measures of nitrous gas into 48 of nitrous oxide. These bodies have an affinity for oxygen ; and the moment they take an atom of oxygen from one of nitrous gas, the atom of azote joins to another of nitrous gas, and forms one of nitrous oxide. In this way, all the azote remains in the nitrous oxide, and just one half of the oxygen. By making the calculation from the preceding table, (page 331) and from the known specific gravities of these gases, it appears that 100 measures of nitrous gas should make 48.5 measures of nitrous oxide, and allow 28.3 measures of oxygen to combine with the bodies introduced. It is very remarkable that these numerical relations should have so long escaped observation.

Sulphuretted hydrogen and moistened iron filings also convert nitrous gas into nitrous oxide : but some ammonia is produced at the expence of the azote, and consequently less nitrous oxide : Davy finds about 42 or 44 per cent.

2. *Nitrous Oxide.*

The gas now denominated nitrous oxide, was discovered, and several of its properties pointed out, by Priestley : he called it *dephlogisticated nitrous gas.* The Dutch chemists published an essay on the subject in the Journal de Physique for 1793, in which the constitution and properties of the gas were more fully investigated. In 1800, Mr. Davy published his Researches, containing a much more complete and accurate developement of the nature of this gas, than had previously been given, as well as of the other compounds of azote and oxygen, and several other collateral ones.

Nitrous oxide gas may be obtained from a salt called *nitrate of ammonia*, being a compound of nitric acid, ammonia and water. The salt is put into a gas bottle, and heat applied, which first fuses the salt, about 300°; by continuing the heat, the fluid salt boils, and is decomposed about 400°, emitting nitrous oxide gas and steam, into which the whole of the salt is principally resolved. The gas may be received either over water or mercury.

The constitution of the salt, nitrate of ammonia, according to Davy, is when crystallized, 18.4 ammonia, and 81.6 acid and water: Now, if we suppose an atom of ammonia to be constituted of one of azote, 5.1, and one of hydrogen, 1, as will be shewn hereafter, and that an atom of the nitrate is composed of 1 atom of each of the elements, ammonia, nitric acid and water, (see plate 4, fig. 36);

we shall have, 6.1 + 19.1 + 8 = 33.2 for the weight of an atom of the salt. This gives 18.4 ammonia, and 81.6 acid and water, exactly agreeing with the experimental results of Davy. The decomposition of an atom of the salt will be found to give one atom of nitrous oxide, weighing 17.2, and two atoms of water, weighing 16. Whence, 100 grains of the salt should be resolved by heat into 51.8 grains of nitrous oxide, and 48.2 grains of water. Mr. Davy decomposed 100 parts of a dried nitrate, that is, one which had lost 8 per cent. of its water of crystallization, and obtained 54.4 nitrous oxide, 4.3 nitric acid, and 41.3 water. Here, as might be expected, the nitrous oxide exceeds, and the water falls short of the calculation, but as nearly as possible in the due proportion. Thus it appears, that whether we consider the genesis of nitrous oxide from the nitrate of ammonia, or from nitrous gas (page 338), still its constitution must be 2 atoms of azote and 1 of oxygen.

The specific gravity of this gas is 1.614; the weight of its atom 17.2 of hydrogen; the diameter in an elastic state (to hydrogen 1) is .947; if a measure of hydrogen contain 1000 atoms, one of nitrous oxide will contain 1176 Most combustible bodies burn in nitrous oxide more vigorously than in common air; it is unfit for respiration, but does not so immediately prove fatal as Dr. Priestley and the Dutch chemists concluded. Mr. Davy found that it may be respired for two or three minutes; and that it generally produces sensations analogous to those of intoxication. It is absorbed by water to the amount of about 80 per cent. according to my recent trials. Davy makes it only 54 per cent., but he was not aware that the quantity is increased in proportion to the purity of the residuary gas. Dr. Henry finds from 78 to 86 per cent. This gas of course expels other gases from water, and is itself driven off unchanged by heat. It is a remarkable fact, that water should take so nearly, and yet not exactly, its bulk of this gas.

Nitrous oxide, by long electrification, loses about 10 per cent. of its bulk; some nitric acid is formed, and a mixture of azote and oxygen is found in the residuum; but no satisfactory decomposition is obtained this way.

All the combustible gases, mixed with nitrous oxide, explode by an electric spark.

Nitrous oxide can be made to combine with the fixed alkalies; but the nature of the compounds has not been much examined.

3. *Nitric Acid.*

Nitric acid, formerly distinguished by the names of *aqua fortis*, and *spirit of nitre*, has been known for three or four centuries. It is now usually procured by distilling a mixture of *nitrate of potash* (saltpetre or nitre) and sulphuric acid. Two parts of the salt by weight, and one of concentrated acid,* are to be mixed in a glass retort; heat is applied, the mixture becomes liquid, and soon exhibits the appearance of ebullition, when a yellowish liquid drops from the retort into a glass receiver. It is nitric acid, one of the most active and corrosive of all the acids. When thus obtained, it is usually pure enough for the purposes of the arts; but it mostly contains both sulphuric and muriatic acid: the former is derived from the acid employed being in part distilled, especially if an excess of it be used and the heat be great; the latter is

* Authors differ greatly as to the proportion of salt and acid: some say 3 salt to 1 of acid: others say nearly equal weights; but 1 acid to 2 salt is that which will nearly saturate the base, and must therefore be right, unless an excess of sulphuric acid be expedient to displace the nitric, which does not appear.

derived from the nitre, which usually contains some muriates mixed with it. To obtain the acid pure, the nitre should be repeatedly dissolved in warm water, and crystallized, taking out the first formed crystals for use; and the acid, when obtained, should be treated with nitrate of barytes to precipitate the sulphuric acid, and nitrate of silver to precipitate the muriatic acid.

The theory of this process is well understood: nitrate of potash is a compound of nitric acid and potash; sulphuric acid has a stronger affinity for potash than nitric; it therefore displaces the nitric, which with the water of the sulphuric acid and that of the nitre, is distilled by the heat, and the compound of acid and water constitutes the liquid nitric acid above. Near the end of the process, the heat is advanced to 500° and upwards, and the acid is partly decomposed; some oxygen is given out, and nitrous gas, which combines with the acid, and forms *nitrous* acid vapour. This acid becomes mixed with the nitric, and renders it more fuming and volatile. The nitrous acid may be driven from the liquid nitric by heat, and then the last becomes less volatile, and colourless like water.

The specific gravity of the liquid nitric acid thus obtained, is usually from 1.4 to 1.5: By fusing the nitre previously, and boiling the sulphuric acid till its temperature was 600°, I obtained a quantity of acid of 1.52. By re-distilling with a moderate heat, it may be obtained of 1.55, and even as high as 1.62, according to Proust (Journal de Physique, 1799). The strength of the acid, that is, the quantity of real acid in a given weight of the liquid, increases in some proportion with the specific gravity, as will presently be shewn.

Some of the more remarkable properties of the liquid nitric acid follow: 1. It emits white vapour when exposed to the atmosphere, owing to its combination with steam or aqueous vapour: this is rendered more evident in the distillation of nitric acid; if the elastic vapour of the acid is escaping from the receiver, it exhibits a white cloud when breathed upon. 2. It is sour to the taste, when diluted with water. 3. It corrodes animal and vegetable substances, and stains them yellow. 4. It combines with water, and, when concentrated, attracts it from the atmosphere; heat is produced, and a small increase of density. With snow it produces a great degree of cold, and instant liquefaction. 5. It is said to be decomposed by the solar light, giving out oxygen, and becoming orange coloured. 6. It inflames several combustibles, such as very dry charcoal, essential oils, &c. 7. When distilled over sulphur, it converts the sulphur into sulphuric acid. 8. It oxidizes the metals, as has been observed, and gives out nitrous gas. 9. When the vapour of nitric acid is passed through a red hot earthen tube, the acid is decomposed into oxygen and azote. The same decomposition is effected by heating nitre red hot in an iron or earthenware retort. 10. It unites to the alkalies, earths, and metallic oxides, forming salts denominated *nitrates*.

One of the most important considerations relative to nitric acid is the determination of the quantity of real acid in a watery solution of a given specific gravity. This subject has engaged the attention of several eminent chemists, particularly Kirwan, Davy, and Berthollet. Their results are widely different. For instance; in an acid of 1.298 sp. gravity, Kirwan says the real acid is 36¾ per cent. Davy says 48, and Berthollet 32 or 33. (See Journal de Physique, March 1807).* My experience in regard to this particular has

* Berthollet, by mistake, makes Davy represent the acid in question to contain 54 per cent. of acid; but it is the *water* which he says is 54 per cent. and the acid 46, when the sp. gravity is 1.283; so that the difference, great as it is, is not quite so enormous.

been considerable, and I shall now state it briefly

Nitric acid has been stated, on the authority of Bergman, to boil at 248°. This is true, if it relate to acid of the strength 1.42; but to acids of no other strength; in fact, it is the highest possible boiling point of the liquid acid: but if the acid be stronger or weaker, then the farther it deviates from 1.42, the less is the temperature at which it boils. The weakest possible acid must evidently boil at 212°; but the point at which the strongest acid boils has not been determined; it will be found, in all probability, little above the common temperature of the atmosphere: an acid of 1.52, I find, boils about 180 or 185°. Proust's acid of 1.62 would probably boil about 100°, or about the same degree as ether. The results of my experience will be noted more particularly in the following table. Besides this variable temperature of ebullition, there is another concomitant circumstance, which has been hinted at by others: In the Paris Memoirs for 1781, Lassone and Cornette had ascertained that when weak nitric acid is boiled or distilled, the weakest portion comes over first; but when the acid is concentrated, the strongest portion comes over first: In the Irish Transactions, vol. 4, Dr. R.

Percival has noticed some results in the distillation of nitre; 2 lbs. of nitre and 1 of concentrated sulphuric acid were mixed and distilled; the products were received in 3 portions; the first was of the strength 1.494; the second, 1.485; the third, 1.442: Proust, in the Journal de Physique, 1799, relates that he obtained an acid 1.52; this being again distilled, gave for the first product 1.51; for the second, 1.51, nearly colourless, which he expected indicated a superior specific gravity; but what surprised him more, was to find the residue colourless, and 1.47. This residue was distilled; the first portion was 1.49, and the rest 1.44. In another instance an acid 1.55 was obtained; this redistilled gave, first 1.62, the second 1.53, and the residue was 1.49.—From all these facts, it appeared to me reasonable to conclude that an acid of some one strength, and only one, was incapable of any change of strength by distillation; or was of such a nature, that the distilled part and the residue were always of the same strength and specific gravity. The actual strength of this acid was a desirable attainment; for such an acid evidently marks a nice adjustment of affinities between the acid and water; or a kind of mutual saturation of the two. By repeated experiments I find this acid to be of the specific gravity 1.42; it is remarkable also that this strength is that which has the boiling temperature a maximum, or 248°. Any acid of inferior strength, being distilled, the weakest part comes over first; and, *vice versâ*, with one of superior strength. For instances, by distilling part of an acid of 1.30, I found an acid of 1.25 in the receiver: again, 530 measures of acid, 1.43 were subjected to distillation; 173 measures were drawn over of 1.433, and 354 of 1.427 were left in the retort: again, by boiling an acid of 1.35 for some time, it became 1.39; and another of 1.48 became 1.46: in short, the continued boiling of any acid, weak or strong, makes it approach more and more to the density 1.42, and to the temperature 248°.

With respect to the quantity of real acid in a solution of given specific gravity, I find it thus: Agreeably to the experience of Kirwan, Richter, Davy, and my own, I conclude that fused nitre is constituted nearly of 47.5 pure acid, and 52.5 potash per cent. Having dissolved 25 parts of this nitre in 100 water, I find the specific gravity, at 60°, = 1.130, and consequently 110.6 measures of the solution. Any given nitric acid is saturated with pure carbonate of potash, and reduced to the specific gravity of 1.130; the measure of the solution is then found, and hence we have data to calculate the real acid in the said solution. Now, 106 grains of 1.51 nitric acid + 248 grains of a solution of potash 1.482, with water, gave 665 grain measures of solution of nitre of 1.130 sp. gravity, indicating 150 of pure nitre. Hence 106 grains of the acid contained 71.2, or 67 per cent. which is $1\frac{1}{2}$ per cent. less than Kirwan deduces it; and this may partly arise from the escape of some acid by its mixture with water producing heat. Again, 133 grains of 1.42 acid were saturated with potash; they gave 672 measures of 1.13 solution, indicating 152 nitre; hence 133 acid contained 72 real, or 54 per cent. which nearly agrees with Kirwan's. Again, 205 grains of 1.35 acid were saturated with 290 grains of 1.48 carbonate of potash; this diluted gave 850 measures of 1.13 solution, indicating 192 nitre; that is, 205 grains acid contained 91 real, 44.4 per cent. which also nearly agrees with Kirwan. Again, 224 grains of 1.315 acid, took 300 grains of 1.458 carbonate of potash; this diluted gave 804 measures of 1.13 solution, indicating 192 nitre; that is, 224 grains of acid contained 86.5 real, = 38.6 per cent.; this is extremely near Kirwan's estimate.

Being thus satisfied with the near approximation to truth of Kirwan's table of nitric acid,

I was notwithstanding desirous to discover, if possible, the sources of error which have influenced the the conclusions of Davy and Berthollet on this subject, whose results are so different from each other and from those of Kirwan.

That Mr. Davy has overrated the quantity of real acid in different solutions is manifest from this ; he finds the acid 1.504 to contain 91.5 per cent. ; now, according to this, an acid of 1.55 would be nearly pure or free from water ; whereas nitric acid has been obtained of the specific gravity 1.62, without there being any reason to suppose it was free from water. Mr. Davy's method of combining the elastic fluids nitrous gas and oxygen, in order to form nitric acid pure and free from water in the first instance, and then combining the acid with a given portion of water, was certainly highly ingenious, and it seems to have been executed with great care ; but that the results this way cannot be relied on, I am convinced from my own experience, some account of which will presently be given. But what appears most surprising and unaccountable in his results, is, how the combination of 47.3 parts of *his* acid with 52.7 parts of potash should form nitre. He relates two experiments ; in one, 54 grains of 1.301 acid combined with potash, gave 66 grains of nitre, at 212°, and this became 60 by fusion : in the other, 90 grains of 1.504 acid, saturated with potash, gave 173 of dry nitre.—In all the similar experiments which I have made, I have uniformly found only *three quarters* of the quantity of nitre said to have been obtained above, from given quantities of the acid. I conclude, therefore, that Mr. Davy must have committed some oversight in these two experiments, and that the direct formation of nitre from nitric acid and potash, accords only with Kirwan's estimate of the strength of nitric acid.

Berthollet, in the Journal de Physique, March 1807, informs us, that he saturated 100 parts of potash with nitric acid of 1.2978 strength, and obtained 170 parts of nitre ; he calculates the acid to contain 32.41 per cent. real, by which we may infer that 216 grains of it were required. Nitre, according to this, would be 100 potash + 70 nitric acid, or 59 potash + 41 acid per cent. This is much more potash than ever before was detected in nitre. How are we to be satisfied that the potash used contained no water ? If it contained any water, this would disappear in the process, and its weight be supplied by nitric acid, which would not be placed to the acid's account. That this was the real fact I have no doubt; 170 parts nitre are constituted of about 89 potash, and 81 nitric acid; the supposed 100 parts of potash were, I conceive, 89 parts potash and 11 water, which of course caused the acid to be underrated by 11 parts.* To prove this, we have only to take a quantity of carbonate of potash, such as is known to contain 89 parts potash; for instance, 170 parts of the dry neutralized carbonate, or 200 crystallized, which Berthollet rightly determines to contain 89 parts of potash, and to this add 216 parts of the above nitric acid, and 170 nitre will be formed. This will also establish another fact worthy of notice; namely, that the quantities of nitric and carbonic acid are the same to a given weight of potash.

I shall now proceed to give the table of the strength of nitric acid. I have copied Kirwan for the strength due to each specific gravity,

* Since writing the above, I have been favoured with the receipt of "*Memoires de Physique et de Chimie de la Société d'Arcueil.* Tome 2." In this there is, amongst other very important and valuable papers, one on the proportion of the elements of some combinations, by Berthollet. The author there determines, page 53, that potash kept for some time in fusion, still retains between 13 and 14 per cent. of water. Hence, he admits the strength of nitric acid above given as his to be erroneous. In the sequel, he concludes that fused nitrate of potash contains 51.4 potash and 48.6 nitric acid.

except the first and second column, which his table has not, and the three last, where I think he has overrated the quantity of acid ; indeed, the lower part of his table is confessedly less correct. I have already given my reasons for considering his table as approximating nearest to the truth ; but have no doubt it might be made more correct ; I have, therefore, only extended the table to two places of decimals in the column of specific gravity. The columr of acid per cent. by measure, will be found convenient for the practical chemist. The first column shews the number of atoms of acid and water in combination or collocation in each solution, agreeably to the preceding determinations ; namely, an atom of acid is taken as 19.1 by weight, and an atom of water as 8. The last column exhibits the boiling points of the several solutions, as found by experiment. Those who wish to repeat these experiments, may be informed that a small globular glass receiver, of the capacity of 6 or 7 cubic inches was used, 2 or 3 cubic inches of acid were put in, and then a loose stopper. It was then suspended over a charcoal fire. When signs of ebullition began to appear, the stopper was withdrawn, and a thermometer, previously adjusted at the boiling point of water, was inserted. It may be proper to observe, that acids which have not previously been boiled, or which contain nitrous acid, usually begin to boil below 212° ; but the vapour soon escapes, and the temperature advances to a stationary point. Nitric acid varies in specific gravity by temperature more than any other, as may be seen, page 44 ; there is, however, an error of the press in the table alluded to, for alcohol and nitric acid ; the numbers should be .11, and not .011. Every 10° counts 6 upon the third place of decimals ; that is, if an acid be 1.516 at 50°, it will be 1.51 at 60°. The expansion with me is uniform, and not variable as with Kirwan.

Table of the quantity of real acid in 100 parts of liquid nitric acid, at the temperature of 60°.

Atoms. Acid. Water.	Acid per cent. by weight.	Acid per cent by measure.	Specific gravity.	Boiling point.
1 + 0	100	175 ?	1.75 ?	30° ?
2 + 1	82.7	134	1.62	100° ?
1 + 1	72.5	112	1.54	175°
	68	102	1.50	210°
	58.4	84.7	1.45	240°
1 + 2	54.4	77.2	1.42	248°
	51.2	71.7	1.40	247°
1 + 3	44.3	59.8	1.35	242°
1 + 4	37.4	48.6	1.30	236°
1 + 5	32.3	40.7	1.26	232°
1 + 6	28.5	34.8	1.22	229°
1 + 7	25.4	30.5	1.20	226°
1 + 8	23	27.1	1.18	223°
1 + 9	21	24.6	1.17	221°
1 + 10	19.3	22.4	1.16	220°
1 + 11	17.8	20.5	1.15	219°
1 + 12	16.6	18.9	1.14	219°

Remarks on the above Table.

1. It seems not improbable, but that an acid free from water may be obtained, as represented in the first line of the table. That such an acid would be in the liquid state, but with a strong elastic steam or vapour over it, at the common temperature, is most probable; in this respect it would resemble ether, but perhaps be more volatile. Seventeen per cent of water would bring it down to acid of the second line, and such as has actually been obtained by Proust. This last would nearly agree with ether in volatility. With respect to the specific gravity of pure nitric acid, it must be less than 1.8; because a measure of that sp. gravity mixed with a measure of water, would make 2 measures of 1.4, *if there were no increase of density;* and acid of this density is nearly half water.* I apprehend if

* The theorem for specific gravities is $\frac{H}{S}+\frac{L}{s}=\frac{H+L}{f}$, where H represents the weight of the body of greatest specific gravity, S its specific gravity, L the body of least specific gravity, *s* its specific gravity, and *f* that of the mixture or compound. Hence in the case above, $\frac{1.8}{1.8}+\frac{1}{1}=\frac{2.8}{1.4}$.

acid of the second line were distilled by a very gentle heat, when mixed with the strongest sulphuric acid, that probably an acid free from water would come over; at least, a concentration is effected by such process in other cases of weaker acids. The receiver should be surrounded with a cold mixture. By distilling an acid 1.31 off sulphuric acid, I got an acid 1.43; and an acid of 1.427 treated in the same manner, gave an acid of 1.5.

2. The acid in the second line, consisting of 2 atoms of acid and 1 of water, having only been obtained by one person, and not particularly examined, we know of no peculiar properties it has, besides the specific gravity and boiling temperature; but there can be little doubt that it possesses other properties, which would distinguish it from all other acids.

3. The acid in the third line, consisting of 1 atom of acid and 1 of water, has not often been obtained, and is therefore little known; it seems to be that acid which fused nitre, and the strongest possible sulphuric acid (such as it is to be had by that mode of concentration, which consists in boiling the common acid) would give by distillation. The water in this case, I suppose, is derived from the sulphuric acid, not from the nitre. It may, however,

be obtained by repeated distillations of any acid above 1.42; provided there is a sufficient quantity of that, and the first products always taken. What the distinguishing properties of this acid may be, I have not had an opportunity of investigating.

4. The acid which consists of 1 atom of acid and 2 of water, is possessed of striking peculiarities. It is in fact that which constitutes a complete reciprocal saturation of the two elements. Evaporation produces no change in its constitution; it distills as water, or any other simple liquid does, without any alteration. It acquires the temperature 248° at boiling, which is greater than any other compound of the two elements acquirés. At any strength above this, the acid is most copiously elevated by heat; at any strength below, the water is most easily raised. Pure water boils at 212°; pure acid perhaps at 30°; the union of both produces a heavier atom than either, and requires a higher temperature for ebullition; but in proportion as either principle prevails more than is necessary for saturation, then the temperature at ebullition is reduced towards that of the pure element itself. Proust has observed that nitric acid of 1.48, produces no more effervescence with tin than with sand; whereas the lower acids act

most violently, as is well known. The fact I find as Proust states it. This would lead one to think that acid of 1.48 was of some peculiar constitution; but I presume this characteristic of nitric acid belongs to that of 1.42, rather than 1.48: not but that the former certainly acts on tin; but the explanation I conceive is this; when the nitric acid in its action on metals is disposed to form ammonia, (an element constituted of one atom of azote and one of hydrogen united) 1 atom of nitric acid and 1 of water are decomposed; the 3 atoms of oxygen go to the metal, and the azote and hydrogen unite and form an atom of ammonia; if, therefore, there were 1 atom of acid to 2 of water, there could be 1 atom of water detached, which would of course join to the remaining acid, and dilute it the more; but if there were 2 atoms of acid for 3 of water, then, detaching 3 atoms of oxygen, would leave an atom of nitrate of ammonia and 1 of water, constituting the salt of that name, and one surplus atom of water. In this case, the remaining acid is not diluted with water by the process, lower than 1 to 2. Such acid, therefore, (which is about 1.47) is probably the lowest that can operate upon tin this way without any effervescence.

5. The acid composed of 1 to 3 water, has not any peculiarity yet discovered.

6. The acid of 1 to 4 water, is remarkable for being that which freezes the most easily of all, namely at — 2° of Fahrenheit, according to Cavendish. The strength of the acid is such, as that 1000 parts dissolve 418 of marble: Now, 418 of marble contain 228 of lime, and these require 370 or 380 of nitric acid, which therefore agrees with the acid of 1 to 4 water, and with that only. Above that strength, or below, the acid requires a greater cold to freeze it.—The inferior acids appear to have no remarkable differences, except such as the table shews ; but the temperature of freezing descends to some undetermined point, and then ascends again.

7 The notion of those who consider the intensity of acid solutions to be proportionate to the quantity per cent. of the acid, or to their density, seems incorrect as far as nitric acid is to determine. It is true, the acidity or *sourness* of the solution, the power to produce effervescence with carbonates, and perhaps other properties, increase nearly as the quantity or strength ; but the freezing and boiling temperatures, the action on metals, as tin, &c. have successive waves, and abrupt termi

nations, which indicate something very different from that gradation in action which varies in the ratio of the quantity.

I have frequently attempted to exhibit the nitric acid in a pure elastic form, and free from water, but have uniformly failed. Some account of the experiments may, notwithstanding, have its use. In order to form the nitric acid free from nitrous and oxynitric, I used large receivers and quantities of gas, amounting to some hundreds of cubic inches, and delivered the nitrous gas to the oxygen, and *vice versa,* in the centre of the receiver, and slowly: still the ratio of oxygen to nitrous gas was variable. The experiments were made over water. Wishing to exclude water as much as possible, I procured some globular receivers, containing from 15 to 60 cubic inches; to these stopcocks were adapted, so as to connect them with the air-pump or with other receivers. These were first filled with oxygen gas or common air, and then partially exhausted; afterwards they were connected with receivers over water, containing known quantities of nitrous gas, and a communication opened; the moment after the nitrous gas had entered the globe, the cock was turned; great care was taken to dry the globe previously to the experiment, and to prevent any water entering with the air, (except the steam which gases commonly have, the quantity of which is easily ascertained for any temperature) The instant the two gases were mixed, the globe was filled with dense orange coloured gas, which continued without any change ; a dewy appearance on the inside of the glass was always perceived, consisting, no doubt, of condensed acid and water.

The results of the experiments are below :

	oxygen		nitrous gas.		per cent.
1.—	1 measure	took	1.8,	residuary	13.6 oxyg.
2.—	1 ——	——	2.11	——	6. nitrous
3.—	1 ——	——	1.44	——	27. oxyg.
4.—	1 ——	——	1.83	——	4. ——
5.—	1 ——	——	2.29	——	2.5 nitrous
6.—	1 ——	——	1.61	——	7.6 oxyg.
7.—	1 ——	——	1.65	——	9.3 nitrous
8.—	1 ——	——	1.8	——	2.5 oxyg.

The residuary gas was examined after letting in water, and washing away the acid. From these results, it is evident the quantity of nitrous gas combining with a given volume of oxygen in such circumstances, is extremely variable, and much like what takes place in small quantities in tubes. The coloured gas is always, I apprehend, either nitrous or oxynitric acid; the nitric acid vapour is without colour, and condenses along with the steam on the sides of the vessel; but the other acids instantly colour the liquid. By inclosing a manometer, I endeavoured to find the elastic force, and the specific gravity of the aerial acids; but from the liquid condensation of a part, I found the specific gravity variable, and always too much. It was commonly about three times that of atmospheric air. Mr. Davy combined 1 measure of oxygen with 2.32 of nitrous gas, leaving an excess of oxygen, and calculated the specific gravity of the aerial product at 2.44; but it is more than probable that this is overrated for the reasons just mentioned. Reasoning by analogy, nitric acid gas should be of the same weight as carbonic acid gas, as its atom is of the same weight; or about the same as nitrous oxide and muriatic acid; hence we may infer, till it can be ascertained experimentally, that the specific gravity of pure nitric acid, in the elastic state, is between 1.5 and 2. Nitrous acid is probably about 2.5, and oxynitric about 2 or 2.25.

I was in hopes to ascertain the constitution of nitric acid, by decomposing nitre by heat, and finding the ratio of azote to oxygen; but, as has been observed by others, the air is of different qualities at different periods of the decomposition. By one experiment, I obtained about 30 grains of air from 100 of nitre in an iron retort; it was received in 5 portions: the first contained 70 per cent. of oxygen, agreeing with the constitution of nitric acid exhibited in the table, page 331; but the succeeding portions gradually fell off, and the last contained only 50 per cent. oxygen.

It may be proper to remark, that the nitric acid of commerce is sold under the names of double and single *aqua fortis*; the former is intended to be twice the strength of the latter; the absolute strength of double aqua fortis is not, I believe, uniform. It commonly runs between the specific gravities of 1.3 and 1.4.

4. *Oxynitric Acid.*

The existence of oxynitric acid is inferred from the combination of oxygen and nitrous gas, in the second experiment, page 328; at least an acid product is obtained, containing more oxygen than is found in nitric acid. As yet I have not been able to obtain this acid any other way, and therefore have not had an opportunity of examining its properties, except upon a very small scale. I thought that distilling the common nitric acid from the oxide of manganese might afford an acid more highly oxydized; but I obtained a product yielding the fumes of oxymuriatic acid, owing no doubt to the muriatic acid previously in the nitric; for, by boiling, these fumes vanished, and left nothing but nitric acid, as far as appeared. The acid obtained from the gases abovementioned, is only at best one half oxynitric, and the other half nitric, so that it is still but a mixture.

A dilute solution of the acid obtained by mixing nitrous and oxygen gas as above, seems to possess similar properties to nitric acid solutions. It is acid to the taste, changes vegetable blue to red, and neutralizes the alkalies; whether in this last case it parts with its excess of oxygen, I have not determined. The atom of oxynitric acid must, it is presumed, weigh 26.1; it consists of 1 atom of azote and 3 of oxygen. The specific gravity of the acid in an elastic state is probably about 2 or $2\frac{1}{4}$.

5. *Nitrous Acid.*

The compound denominated nitrous acid is obtained by impregnating liquid nitric acid with nitrous gas. This acid, however, is never pure nitrous acid, but a mixture of nitric and nitrous; as is evident by boiling it, when the nitrous is driven off, and the nitric remains behind. Pure nitrous acid seems to be obtained by impregnating water with oxygenous gas, and then with nitrous gas; in this way 1 measure of oxygen takes about $3\frac{1}{2}$ of nitrous; that is, 1 atom of oxygen takes 2 atoms of nitrous gas to form 1 of nitrous acid. The weight of the atom therefore is 31.2.

By repeated trials I find that 100 measures of nitric acid of 1.30 specific gravity, agitated with nitrous gas, takes up about 20 times its bulk of the gas. If the acid be of twice the strength, or of half the strength, it makes little difference; the quantity of gas is nearly as the real acid, within certain limits of specific gravity. Very dilute acid (as 1 to 300 water) seems to have scarcely any power of absorbing nitrous gas, besides what the water itself has. Hence, it seems that what we call nitrous acid, is only about $\frac{1}{12}$th of it real acid; the rest is nitric acid.

Mr. Davy concludes, that the bright yellow acid of 1.50 specific gravity, contains nearly 3 per cent. of nitrous gas; the dark orange $5\frac{1}{2}$, and the blue green 8; the two last being of the strength 1.48 or 1.47.

From the experiments of Priestley, it is evident that the nitrous acid, or as he called it, the *phlogisticated nitrous vapour*, is much more volatile than nitric acid; or, to speak more properly, has less affinity for water. Hence the fuming of the nitrous acids in great part arises. This is further corroborated by the ready ebullition of those acids. The acid which I obtained above by saturating nitric acid of 1.30 with nitrous gas, was dark orange, and strongly fuming: it boiled at 160°; whereas the nitric acid of the same strength boils at 236°. It is owing to the same cause that very dilute nitrous acid exhibits the characteristic smell of the acid; but equally dilute nitric acid has no smell. When nitrous acid is diluted so far as to contain just its own bulk of nitrous gas, it then attracts oxygen, but very slowly; it requires as much agitation as sulphuret of lime to saturate it.

It does not appear that pure nitrous acid

combines with the alkalies so as to form dry salts or *nitrites*; the concentrated solutions seem to lose the nitrous gas, and then the *nitrates* are obtained.

SECTION 3.

OXYGEN WITH CARBONE.

There are two compounds of oxygen and carbone, both elastic fluids; the one goes by the name of *carbonic acid*, the other *carbonic oxide*; and it appears by the most accurate analyses, that the oxygen in the former is just double what it is in the latter for a given weight of carbone. Hence, we infer that one is a binary, and the other a ternary compound; but it must be enquired which of the two is the binary, before we can proceed according to system. The weight of an atom of carbone or charcoal, has not yet been investigated. Of the two compounds, carbonic acid is that which has been longest known, and the proportion of its elements more generally investigated. It consists of nearly 28 parts of charcoal by weight, united to 72 of oxygen. Now, as the weight of an atom of oxygen has been determined already to be 7; we shall have the weight of an atom of carbone = 2.7, supposing carbonic acid a binary compound; but 5.4, if we suppose it a ternary compound.

Carbonic acid is of greater specific gravity than carbonic oxide; and on that account, it may be presumed to be the ternary or more complex element. It must, however, be allowed, that this circumstance is rather an indication than a proof of the fact. The element of charcoal may be so light, that two atoms of it with one of oxygen, may be specifically lighter than one with one. But there are certain considerations which incline us to believe, that the element of charcoal is not much inferior to oxygen in weight. Oils, alcohol, ether, wood, &c. are compounds into which hydrogen and charcoa principally enter; these are a little lighter than water, a compound of hydrogen and oxygen. Though charcoal in a state of extreme division is readily sublimed by heat, it does not assume the form of a permanently elastic fluid, which one would expect of a very light element. Besides, carbonic acid is the highest degree of oxidation of which charcoal is susceptible, as far as we know; this rarely happens under two atoms of oxygen. Carbonic acid is easily resolved by electric shocks into oxygen and carbonic oxide; but carbonic oxide does not appear to be resolved in the same mode into charcoal and carbonic acid, which one might expect from a triple compound. One of the most common ways of obtaining carbonic oxide, is to decompose carbonic acid by some substance possessing affinity for oxygen; now, oxygen may be abstracted from a body possessing two atoms of it more easily than from one possessing only one. On all these accounts, there can scarcely be a doubt that carbonic oxide is a binary, and carbonic acid a ternary compound.

1. *Carbonic Oxide.*

This gas was discovered by Dr. Priestley; but its distinguishing features were more fully pointed out by Mr. Cruickshanks, in an essay in Nicholson's Journal, 1801. About the same time, another essay of Desormes and Clement was published in the Annales de Chemie, on the same subject. These essays are both of great merit, and highly creditable to their authors. Before that time, carbonic oxide had been confounded with the combustible gases composed of carbone and hydrogen;

but Cruickshanks and Desormes distinctly demonstrated, that in the combustion of this gas nothing but carbonic acid was produced ; and that the quantity of oxygen requisite for its combustion, was not more than half of that afterwards contained in the carbonic acid ; they, therefore, rightly concluded that the gas was a compound of carbone and oxygen, since which it has been known by the name of carbonic oxide.

Carbonic oxide may be procured by various processes ; but it is mostly accompanied with one or more foreign gases, from some of which it is difficult to separate it ; for this reason, when it is wanted pure, such methods must be used as give it mixed with gas that can be extracted. The following process answers well : Let equal weights of clean, dry iron filings and pulverized dry chalk, be mixed together, and put into an iron retort ; let the retort be heated red, and the heat gradually increased ; gas will come over copiously, which may be received over water ; this gas will be found a mixture of perhaps equal parts of carbonic oxide and carbonic acid ; the last may be extracted by due agitation in a mixture of lime and water ; what remains is pure carbonic oxide, except 2 or 3 per cent. of common air, from the lime water. The theory of this process is manifest; chalk consists of carbonic acid and lime; the carbonic acid is disengaged by heat, and is immediately exposed to the red hot iron, which in that state has a strong affinity for oxygen; the carbonic acid parts with one half of its oxygen to the iron, and the residue is carbonic oxide; but part of the acid escapes along with it undecompounded. With a proper apparatus, the gas may be procured by transmitting carbonic acid repeatedly over red hot charcoal in an iron or porcelain tube.

This gas may be obtained, by exposing to a red heat, a mixture of charcoal with the oxides of several metals, or with carbonate of lime, barytes, &c. But there is great danger in this way of procuring some hydrogen, and carburetted hydrogen, along with carbonic oxide and acid. Indeed, all gas procured from wood and from moist charcoal, is a mixture of these four, varying in proportion according to the heat and the continuance of the process.

According to Cruickshanks, the specific gravity of carbonic oxide is .956; according to Desormes and Clement, .924. Apprehending that they had both rated it too low, I carefully found the specific gravity of a mixture of 6 parts carbonic oxide and 1 common air, at two trials; in one it came out .945, and in the other .94; I conceive, then, that .94 may be taken as a near approximation to the truth; it is just the mean of the two authors above. Carbonic oxide is fatal to animals that breathe it; it is combustible, and burns with a fine, clear, blue flame, without any smoke or the least appearance of dew, if a bell glass is held over the flame. This circumstance, amongst others, distinguishes it clearly from all gases containing hydrogen, either mixed or combined. When mixed with oxygenous gas, or common air, in Volta's eudiometer, it explodes with an electric spark, and is converted into carbonic acid. The circumstances attending the explosion are somewhat remarkable; unless the carbonic oxide amount to at least $\frac{1}{5}$th of the mixture, it will not explode; and the oxygen must be at least $\frac{1}{15}$th of the mixture. Besides, it frequently happens, when common air is used for oxygen, that a smart explosion takes place, and yet both carbonic oxide and oxygen shall be found in the residuum. This circumstance disappears if the oxygen be above 30 per cent. pure. It should be observed, that whenever proportions near the extremes above noted, are used, the results become ambiguous; as a partial combustion sometimes happens. When

100 measures of carbonic oxide are mixed with 250 of common air, (in which case the whole of the combustible gas should combine with the whole of the oxygen) a smart explosion ensues by the first spark; but only $\frac{2}{3}$ds of the gas is burnt; the rest, and a corresponding proportion of oxygen, remain in the residuum. When plenty of combustible gas and a minimum of oxygen are exploded, the whole of the oxygen usually disappears.

Carbonic oxide does not explode by electricity when mixed with oxymuriatic acid, at least in any instance I have had, unless a small portion of common air be present; but the mixture being exposed to the sun, a diminution soon takes place; if the light be powerful, 5 or 10 minutes are sufficient to convert 100 grain measures of the gas along with 100 of the acid, into carbonic and muriatic acids. I have not been able to determine, from the lateness of the season (October), whether the mixture would explode by the solar light.

Pure carbonic oxide is not at all affected by electricity. I was present when Dr. Henry conducted an experiment, in which 35 measures of carbonic oxide received 1100 small shocks; no change of dimensions took place; there was no carbonic acid formed, nor oxygen liberated; but the residuary gas being fired with oxygen appeared to be pure carbonic oxide.

Water absorbs $\frac{1}{27}$th of its bulk of carbonic oxide. It will be seen by reference to page 201, also to the Manchester Memoirs, vol. 1. *new Series*, pages 272 and 436, that this gas has perplexed me more than any other, at different periods, as to what class to refer it, in regard to absorption. One reason was, that in my more early experiments I used sometimes to obtain carbonic oxide by means of charcoal; in which case it was doubtless mixed with more or less of hydrogen; another reason was, that I did not agitate the water long enough; this gas requires longer agitation than any other I have met with. I can now make water take up full $\frac{1}{27}$th of its bulk, or at least in that proportion, according to the purity of the incumbent gas.

The proportion of carbone and oxygen found in carbonic oxide, has been found by experiment as under:

	measures.		measures.		measures.
Cruickshanks —	100carb. ox.	prod.	92carb. ac.	—take	40oxy.
Desormes&Clem.	100	—	79	—	36 —
—— —	—	—	79	—	39 —
—— —	—	—	88	—	34 —
My own exp. —	100	—	94	—	47 —

Cruickshanks certainly underrates the oxygen; I always find the oxygen fully equal to half the carbonic acid, whether fired over mercury or water. Desormes' experiments were made over water, and are therefore rather uncertain as to the quantity of acid; they have evidently used impure gas. Their first result given above is the mean of nine experiments; the other two are extremes in regard to acid and oxygen (Annales de Chimie 39—page 38). It is remarkable, that in one of their deductions (page 44), on which they seem to rely most, they find the carbone 44, and the oxygen 56 parts: by a previous experiment, they had found carbonic acid to consist of 28.1 carbone, and 71.9 oxygen (page 41); that is, of 44 carbone, and 112 oxygen: where the oxygen is just double of that in the carbonic oxide to a given quantity of carbone. This most striking circumstance seems to have wholly escaped their notice.

The exact composition of this gas is easily ascertained by exploding it with common air over water. Let 2 parts of the gas be mixed with 5 of air, and fired; the residuum must be washed in lime water, and the quantity left accurately noted; then apply a small portion of nitrous gas to the residuum, sufficient to take out the oxygen; hence we have data to find the quantity of the two gases which have combined to form carbonic acid. In this way, 10 measures of oxide will be found to take from 4.5 to 5 measures of oxygen.

The conclusion then is, that carbonic oxide in its combustion, requires just as much oxygen as it previously has in its constitution, in order to be converted into carbonic acid. This agrees too with the results derived from the specific gravity of the gas. The gas may be considered as *half burned charcoal*; it bears the same relation to carbonic acid as nitrous gas does to nitric acid. An atom of carbonic oxide consists then of one of carbone or charcoal, weighing 5.4, and one of oxygen. weighing 7, together making 12.4. The diameter of the atom, in an elastic state, is 1.02, that of hydrogen being unity. Or, 106 measures of the gas contain as many atoms as 100 measures of hydrogen.*

* It will, perhaps, be expected that some notice should be taken here of the opinion of Berthollet, that carbonic oxide is a compound of carbone, oxygen, and hydrogen, and therefore may be denominated *oxycarburetted hydrogen.* It was formerly his opinion that certain gases consist of carbone and hydrogen, and hence are called *carburetted hydrogen*; others consist of carbone, oxygen, and hydrogen, and are denominated as above. But in the 2d volume of the Memoirs d'Arcueil, he contends that all the combustible

2. *Carbonic Acid.*

The gas now denominated carbonic acid, has been recognised as an elastic fluid distinct from atmospherical air, for a longer time perhaps than any other. It may be said to have

gases that have been considered as belonging to these two species, are in fact oxycarburetted hydrogen; and that these elements are combined in an indefinite variety of proportions. That the combustible gases produced from moist charcoal and other bodies, contain oxygen, carbone, and hydrogen in various proportions, is a fact of which no experienced person can doubt; but it has not yet been shewn satisfactorily by any one, that they cannot be made by mixing certain proportions of two or more of the following distinct species, namely, *carburetted hydrogen* (of stagnant water), *carbonic oxide, olefiant gas,* and *hydrogen.* —As for carbonic oxide, whilst it remains an indisputed fact, that *in the combustion of it nothing but carbonic acid is produced, and that equal in weight to the carbonic oxide and the oxygen,* it will require very specious reasoning to convince any one that it contains either hydrogen, sulphur, or phosphorus; unless it be first proved that carbonic acid contains the same. One argument of Berthollet is, however, more ingenious than any reply to it which has appeared: it is this, *a compound elastic fluid ought to be found specifically heavier than the lighter of the two elementary fluids constituting it.* This is, as far as I know, universally true; but it does not follow that carbonic oxide should be specifically heavier than oxygenous gas. An atom of char-

been known, though very imperfectly, to the ancients. Towards the close of the last century, almost all the distinguished chemists had occasionally turned their attention to this article, and its properties became gradually developed. It has received at times different names; namely, *choak damp*, *fixed air*, *aerial acid*, *mephitic*, and *calcareous acid.*

coal, it appears, is lighter than an atom of oxygen; it is probable, then, it would make a lighter elastic fluid, could we convert it into one by a due degree of heat. We cannot judge of the specific gravity of an elastic fluid either from the weight of the article in a solid or liquid form; or from the degree of heat requisite to produce the elastic state. Water is certainly heavier than charcoal; yet it produces a light elastic fluid. Ether is lighter than water; but it produces a heavier elastic fluid, and at a lower temperature. Carbonic oxide may be lighter than oxygen, for the same reason that nitrous gas is lighter than oxygen; namely, because oxygen is the heavier of the two elements that enter into its composition. The answers above alluded to deny the generality of the argument; they produce what they conceive a parallel case in nitrous oxide, and nitrous gas; and allege that oxygen, the heavier of the two component elements, being abstracted from nitrous gas, leaves nitrous oxide, which is specifically heavier than nitrous gas. But if the doctrine we have advanced on this head be true, they have mistaken *half* of the operation for the *whole;* in the conversion alluded to, not only the oxygen is taken from an atom of the nitrous gas, but at the same moment the azote is joined to another atom of the nitrous gas to form one of nitrous oxide.

Carbonic acid gas is formed by burning charcoal; but it is most easily obtained in a pure state from chalk, or some of the carbonates, by means of dilute sulphuric or other acid; it may be received in bottles over mercury or water, but the latter absorbs a portion.—This gas extinguishes flame, and is unfit for respiration; its specific gravity is nearly 1.57, as appears from the experience of all who have tried: 100 cubic inches, at the pressure of 30 inches of mercury, and temperature of 60°, weigh from 47 to 48 grains. Carbonic acid is frequently produced in mines, and in deep wells: it is known to workmen by the name of *choak damp*, and proves fatal to many of them; it is also constantly found in the atmosphere, constituting about $\frac{1}{1000}$th part of the whole; its presence is easily detected by lime water, over which it forms a film almost instantly. In the breathing of animals this gas is constantly produced; about 4 per cent. of the air expired by man, is usually carbonic acid, and the atmospheric air inspired loses the same quantity of oxygen.

Water absorbs just its own bulk of carbonic acid gas; that is, the density of the gas in the water after agitation, is the same as the density of the incumbent gas above, and the elasticity of the gas in the water is unimpaired. The water so impregnated has the taste and other properties of an acid. This gas is the product of fermentation, and gives to fermented liquors their brisk and sparkling appearance; but it soon escapes from liquids, if they are exposed to the air.

Carbonic acid combines with alkalies, earths and metallic oxides, and forms with them salts called *carbonates*. Lime water, by agitation with any gas containing carbonic acid, becomes milky, owing to the generation of chalk or carbonate of lime, which is insoluble in water. Hence this water is an elegant test of the presence of carbonic acid.

The constitution of this gas can be shewn both by synthesis and analysis; but more conveniently by the former. The experiments of Lavoisier, Crawford, Desormes and Clement, and more recently those of Allen and Pepys, on the combustion of charcoal in oxygen gas, have left no doubt as to the quantity of the elements in carbonic acid; 28 parts of charcoal by weight unite to 72 of oxygen, to form 100 of carbonic acid, very nearly. In this case too, it is remarkable that the volume of carbonic acid is the same as that of the oxygen entering into its constitution. Tennant has shewn that carbonic acid may be decomposed; by heating phosphorus with carbonate of lime, phosphate of lime and charcoal were obtained.

Carbonic acid is decomposed by electricity into carbonic oxide and oxygen. I assisted Dr. Henry in an experiment by which 52 measures of carbonic acid were made 59 measures by 750 shocks; the gas after being washed became 25 measures; whence these had arisen from the decomposition of 18 measures of acid; these 25 measures consisted of 16 carbonic oxide and 9 oxygen; for, a portion being subjected to nitrous gas, manifested $\frac{1}{3}$d of its bulk to be oxygen; and the rest was fired by an electric spark, and appeared to be almost wholly converted into carbonic acid.

Carbonic acid then appears to be a ternary compound, consisting of one atom of charcoal and two of oxygen; and as their relative weights in the compound are as 28 : 72, we have 36 : 28 : : 7 : 5.4 = the weight of an atom of charcoal; and the weight of an atom of carbonic acid is 19.4 times that of hydrogen. The diameter of an atom of the acid in an elastic state is almost exactly the same as that of hydrogen, and is therefore represented by 1; consequently a given volume of this gas contains the same number of atoms as the same volume of hydrogen.

SECTION 4.

OXYGEN WITH SULPHUR.

Two distinct compounds of oxygen and sulphur have been for some time universally recognized; but there exists a third, the nature and properties of which are yet in a great measure unknown. According to the received principles of nomenclature, the first, denoting the lowest degree of oxidizement of sulphur, may be called *sulphurous oxide*, or the *oxide of sulphur*; the second, denoting a higher degree, *sulphurous acid*; and the third or highest degree known, *sulphuric acid*.

1. *Sulphurous Oxide.*

The existence of oxide of sulphur in a combined state was first observed by Dr. Thomson. By sending oxymuriatic acid in the gaseous state, through a vessel containing flowers of sulphur, he obtained a red liquid, which he denominated *sulphuretted muriatic acid*; but it would have been more properly called *mu-*

riate of sulphur; as its formation is similar to that of muriate of iron, &c. in like circumstances. Now, it has been shewn that oxymuriatic acid is muriatic acid united to oxygen, one atom to one; hence the atom of oxygen oxidizes an atom of sulphur, and the muriatic acid unites to the oxide, forming muriate of sulphur, or more strictly *muriate of oxide of sulphur*. This oxide of sulphur, Dr. Thomson finds, is not easily obtained separate; for when the red liquid is poured into water, the oxide resolves itself into sulphur and sulphuric acid. (Nicholson's Journal, vol. 6—104.)

When sulphuretted hydrogen gas and sulphurous acid gas are mixed over mercury, in the proportion of 6 measures of the former to 5 of the latter, both gases lose their elasticity, and a solid deposit is made on the sides of the tube. The common explanation given of this fact is, that the hydrogen of the one gas unites to the oxygen of the other to form water, and the sulphur of both gases is precipitated. This explanation is not correct; water is indeed formed, as is stated; but the deposition consists of a mixture of two solid bodies, the one sulphur, the other sulphurous oxide: they may be distinguished by their colour; the former is yellow, the latter bluish white; and when they are both thrown into water, the former soon falls down, but the latter remains for a long time suspended in the water, and gives it a milky appearance, which it retains after filtration. It will appear in the sequel, that 5 measures of sulphurous acid contain twice as much oxygen as the hydrogen in 6 measures of sulphuretted hydrogen require ; it follows, therefore, that one half of the oxygen ought still to be found in the precipitate, which accords with the above observation. Again, if water, impregnated with each of the gases, be mixed together till a mutual saturation takes place, or till the smell of neither gas is observed after agitation, a milky liquid is obtained, which may be kept for some weeks without any sensible change or tendence to precipitation. Its taste is bitter and somewhat acid, very different from a mere mixture of sulphur and water. When boiled, sulphur is precipitated, and sulphuric acid is found in the clear liquid. The milkiness of this liquid seems therefore owing to the oxide of sulphur.

It may be proper to remark that the white flowers of sulphur, commonly sold by the druggists, are not the oxide of sulphur. They are obtained by precipitating a solution of sulphuret of lime by sulphuric acid. They consist of 50 per cent. sulphate of lime and 50 of sulphur, in some state of combination with the sulphate; for, the two bodies are not separable by lixiviation.

When sulphur in a watch glass is ignited, then suddenly extinguished, and placed on a stand over water, and covered with a receiver, the sulphur sublimes and fills the receiver with white fumes. On standing for some minutes or an hour, the sulphur gradually subsides, and forms a fine yellow film over the surface of the water. The air in the receiver loses no oxygen by this process. But when sulphur ignited, is placed in the circumstances above-mentioned, it burns with a fine blue flame, emitting some bluish white fumes, scarcely perceptible at first; as the combustion continues these fumes increase, and towards the conclusion, when the oxygen begins to be deficient, they rise up in a copious stream, and fill the receiver so that the stand is scarcely visible. If a portion of the air is passed through water, it still continues white. In the space of an hour the air in the receiver becomes clear; but no traces of sulphur are seen on the surface of the water. The whiteness in this last case does not, therefore, seem to arise from sublimed sulphur, but from the oxide of sulphur, which is formed when there is not oxygen sufficient to form sulphurous acid; this last is known to be a perfectly transparent elastic fluid. Whether the sulphurous oxide in this case is absorbed by the water in that state, or is gradually converted into sulphurous or sulphuric acid, I have not been able yet to determine.

When a solution of sulphuret of lime has been exposed to the air for a few weeks, till it becomes colourless, and sulphur is no longer precipitated, if a little muriatic acid be added to it, the whole becomes milky, and exhales sulphurous acid; after some time sulphur is deposited, and the sulphurous acid vanishes, leaving muriate of lime in solution. This milkiness must be occasioned by sulphurous oxide; for, sulphite of lime, treated in like manner, exhibits no such appearance.

As far, then, as appears, sulphurous oxide is a compound of one atom of sulphur and one of oxygen; it is capable of combining with muriatic, and perhaps other acids; when suspended in water, it gives it a milky appearance and a bitter taste, and the mixture being heated, the oxide is changed into sulphur and sulphuric acid. An atom of sulphur being estimated, from other considerations hereafter to be mentioned, to weigh 13, and one of oxygen weighing 7, it will follow that oxide of sulphur is constituted of 65 sulphur and 35 oxygen per cent.

2. *Sulphurous Acid.*

When sulphur is heated to a certain degree in the open air, it takes fire and burns with a blue flame, producing by its combination with oxygen an elastic fluid of a well known and highly suffocating odour; the fluid is called *sulphurous acid*. Large quantities of this acid are produced by the combustion of sulphur in close chambers, for the purpose of bleaching or whitening flannels and other woollen goods. In this way, however, the acid never constitutes more than 4 or 5 per cent. of the volume of air, and is therefore much too dilute for chemical investigations. It may be obtained nearly pure by the following process: To two parts of mercury by weight put one part of concentrated sulphuric acid in a retort; apply the heat of a lamp, and sulphurous acid gas will be produced, which may be received over mercury. The reason of this is, each atom of mercury receives an atom of oxygen from one of sulphuric acid, and the remainder of the sulphuric atom constitutes one of sulphurous acid, as will be evident from what follows.

Sulphurous acid is unfit for respiration and for combustion: its specific gravity, according to Bergman and Lavoisier, is 2.05 ; according to Kirwan, 2.24 ; by my own trials, it is 2.3. I sent a stream of the gas, after it had passed through a cold vessel connected with the retort, into a flask of common air ; this was afterwards weighed, and the quantity of acid gas then ascertained by water ; it appeared by two trials, agreeing with each other, that 12 ounce measures of the gas weighed 9 grains more than the same quantity of common air, and this last weighed 7 grains nearly —Water absorbs about 20 times its bulk of this gas at a mean temperature, according to my experience ; but some say more, others less. The quantity absorbed, no doubt, will be greater as the temperature is less. Hence, it seems that water has a chemical affinity for the gas ; but the whole of it escapes if long exposed to the air, except a small portion which is converted into sulphuric acid.

When water, impregnated with sulphurous acid, is exposed to oxygen in a tube, the oxygen is slowly imbibed, and sulphuric acid formed. In twelve days, 150 measures of the acid, absorbed by water, took 35 of oxygen, leaving a residuum of oxygen and sulphurous acid. When sulphurous acid gas and oxygen gas are mixed and electrified for an hour over mercury, sulphuric acid is formed ; but I do not find that the proportion of the elements of the acids can in this way be ascertained; for, the mercury becomes oxidized, and consequently liable to form an union with either of the acids.—The two gases also combine, when made to pass through a red hot porcelain tube. Sulphurous acid is said to be decomposed by hydrogen and charcoal at a red heat; sulphur is deposited, and water or carbonic acid formed, according as the case requires. When a measure of oxymuriatic acid gas is put to a measure of sulphurous acid gas, over mercury, the sulphurous acid is converted into sulphuric; but no exact result can be obtained, from the rapid action of the former gas on mercury.

Sulphurous acid oxidizes few of the metals; but it possesses the common properties of acids, and unites with the alkalies, earths, and metallic oxides, forming with them salts denominated *sulphites*.

It remains now to investigate the number and weight of the elements in sulphurous acid. I have made a great number of experiments on the combustion of sulphur in atmospheric air, in various circumstances; but those I more particularly rely upon, were made in a receiver containing 400 cubic inches: it was open at top, and had a brass cap, by means of which an empty bladder could be attached to the receiver, in order to receive the expanding air; a small stand was provided, and a watch glass was placed on it, filled with a known weight of the flowers of sulphur; the whole was placed on the shelf of a pneumatic trough, and as soon as the sulphur was lighted by an ignited body, the receiver was placed over it, with its margin in the water; the combustion was then continued till the blue flame expired; near the conclusion, white fumes arise copiously, and fill the receiver. A small phial was then filled with water, inverted, and carefully pushed up into the receiver to withdraw a portion of air for examination; the receiver was then removed, and the loss of sulphur ascertained. The residuary gas in the phial was fired with hydrogen in Volta's eudiometer. The loss of sulphur at a medium was 7 grains, and the oxygen in the residuary gas was at a medium 16 per cent. or rather more; the weight of oxygen, therefore, which had disappeared, was from 5 to 6 grains. Hence it may be said, that 7 grains of sulphur combined with $5\frac{1}{2}$ of oxygen; but as the white fumes are oxidized inferior to sulphurous acid, it is most probable that sulphur requires its own weight of oxygen nearly to form sulphurous acid. In confirmation of this, it is observable, that no material change of bulk is effected in the gas by the combustion; and this is also remarked in the analogous combustion of charcoal. Thus, then, the specific gravity of sulphurous acid should exhibit a near approximation to twice that of oxygen, as it is found to do above. Now, as it would be contrary to all analogy, to suppose sulphurous acid to consist of 1 atom of sulphur and 1 of oxygen, we must presume upon its being 1 of sulphur and 2 of oxygen; and hence the weight of an atom of sulphur will be 14 times that of hydrogen.

Another and more rigid proof of the constitution of sulphurous acid, we obtain from the combustion of sulphuretted hydrogen in Volta's eudiometer. This compound, it will be shewn, contains exactly its own bulk of hydrogen; the rest is sulphur: Their relative weights, as appears from the specific gravity, must be 1 to 14 nearly; now, when sulphuretted hydrogen is exploded with plenty of oxygen over mercury, the whole of the last mentioned gas is converted into water and sulphurous acid; it is found that 2 measures of the combustible gas combine with 3 measures of oxygen; but 2 measures of hydrogen take 1 measure of oxygen; therefore, the sulphur takes the other 2 measures; that is, the atom of sulphur requires 2 atoms of oxygen for its combustion, and that of hydrogen 1 atom of oxygen; which gives the same constitution as that deduced above for sulphurous acid.

The proportions of sulphur and oxygen in this acid, have been variously stated, mostly wide of the truth. We have one account that gives 85 sulphur and 15 oxygen. Dr. Thomson, in Nicholson's Journal, vol. 6, page 97, gives 68 sulphur and 32 oxygen; but in his Appendix to the 3d edition of his Chemistry, he corrects the numbers to 53 sulphur and 47 oxygen. Desormes and Clement say 59 sulphur and 41 oxygen (ibid. vol. 17—page 42). According to the preceding conclusions, if the atom of sulphur be stated at 14; then the proportion of sulphur to oxygen will be 50 sulphur to 50 oxygen, or equal weights; but if sulphur be denoted by 13, then sulphurous acid will consist of 48 sulphur and 52 oxygen per cent., which numbers I consider as the nearest approximation: the diameter of the elastic atom of sulphurous acid is rather less than that of hydrogen, as appears from the circumstance that 5 measures of the gas saturate 6 measures of sulphuretted hydrogen, which last contain as many atoms as the like measures of hydrogen. On this account, the diameter of an atom of sulphurous acid may be denoted by .95, and the number of atoms in a given volume, to that of hydrogen in the same volume, will be as 6 to 5, or 120 to 100.

3. *Sulphuric Acid.*

The sulphuric acid of commerce, commonly known in this country by the name of *oil of vitriol,* is a transparent liquid of an unctuous feel, of the specific gravity 1.84, and very corrosive; it acts powerfully on animal and vegetable substances, destroying their texture, and mostly turning them black. This acid was formerly obtained from green vitriol (sulphate of iron) by distillation; hence the name *vitriolic acid.* It is now commonly obtained by burning sulphur, mixed with a portion of nitre, (from $\frac{1}{8}$th to $\frac{1}{10}$th of its weight) in leaden chambers; sulphuric acid is formed and drops down into water, which covers the floor of the chambers; this water, when charged sufficiently with acid, is drawn off, and subjected to evaporation till the acid is concentrated in a higher degree; when it is put into glass retorts, and placed in a sand bath; the weaker part of the acid is distilled into receivers, and the others concentrated nearly as much as is possible in the circumstances. The acid in the receivers is again boiled down and treated as before.

Some authors have affected to consider the theory of the formation of sulphuric acid as very obvious ; the nitre, they say, furnishes a part of the oxygen to the sulphur, and the atmosphere supplies the rest. Unfortunately for this explanation, the nitre, if it were all oxygen, would not furnish above $\frac{1}{15}$th of what is wanted ; but nitre is only 35 per cent. oxygen ; it cannot, therefore, supply the sulphur with much more than $\frac{1}{30}$th part of what it wants, if all the oxygen were extricated ; but not more than $\frac{1}{2}$ or $\frac{1}{3}$d of this small portion is disengaged from the potash ; for, the salt becomes a sulphate instead of a nitrate, and retains most of the oxygen it had, or acquires oxygen again from some source. Several well informed manufacturers, aware of the fallacy of the above explanation, have attempted to diminish the nitre (which is an article of great expence to them), or to discard it altogether ; but they find it indispensibly necessary in some portion or other ; for, without it they obtain little but sulphurous acid, which is in great part incondensible, and not the acid they want. The manner in which the nitre operates, for a long time remained an ænigma. At length Desormes and Clement, two French chemists, have solved the difficulty, as may be seen in an excellent essay in the Annal. de Chimie, 1806, or in Nicholson's Journal, vol. 17. These authors shew, that in the combustion of the usual mixture of sulphur and nitre, sulphurous acid is first formed, and nitrous acid or nitrous gas liberated, partly from the heat, and partly perhaps from the action of sulphurous acid ; the nitrous gas or acid becomes the agent in oxidizing the sulphurous acid, by transporting the oxygen of the atmospheric air to it, and then leaving them in union, which constitutes sulphuric acid. The particle of nitrous gas then attaches another of oxygen to itself, and transports it to another atom of sulphurous acid ; and so on till the whole is oxidized. Thus the nitrous acid operates like a ferment, and without it no sulphuric acid would be formed.

This theory of the formation of sulphuric acid has so very imposing an aspect, that it scarcely requires experiment to prove it. It is, however, very easily proved by a direct and elegant experiment. Let 100 measures of sulphurous acid be put into a dry tube over mercury, to which add 60 of oxygen ; let then 10 or 20 measures of nitrous gas be added to the mixture ; in a few seconds, the inside of the tube becomes covered with a crystalline appearance, like hoar frost, and the mixture is reduced to $\frac{1}{3}$d or $\frac{1}{4}$th of its original volume. If now a drop of water be admitted, the crystalline matter is quickly dissolved into the water, sparkling as it enters, and the gases entirely lose their elasticity, except a small residuum of azote and nitrous gas. If the tube is then washed out, the water tastes strongly acid, but has no smell of sulphurous acid. It is evident, that in this process the nitrous gas unites to the oxygen, and transports it to the sulphurous acid, which, receiving it from the nitrous, becomes sulphuric acid. It appears, moreover, that solid sulphuric acid is formed when no water is present; and consequently this is the natural state of sulphuric acid entirely free from water. It must be observed, that if any water in substance is present when the mixture of gases is made, the water seizes the nitrous acid as it is formed, and consequently prevents it oxidizing the sulphurous acid; on the other hand, the presence of water seems necessary in the sequel, to take the new formed sulphuric acid away, in order to facilitate the oxidizement of the remaining sulphurous acid. The oxygen necessary to saturate 100 measures of sulphurous acid seems to be about 50 measures; but it is difficult to ascertain this with precision, because the nitrous gas takes up the superfluous oxygen, and begins to act upon the mercury.

Now, it has been shewn, that sulphurous acid contains nearly its own bulk of oxygen, and is constituted of 1 atom of sulphur and 2 of oxygen; and it appears from the above, that half as much oxygen more, that is, 1 atom, converts it into sulphuric acid: hence, the sulphuric acid atom is constituted of 1 atom of sulphur and 3 of oxygen; and if the atom of sulphur be estimated at 13 in weight, and the 3 of oxygen at 21, the whole compound atom will weigh 34 times the weight of an atom of hydrogen; that is, pure sulphuric acid consists of 38 sulphur and 62 oxygen per cent.

In the year 1806, by a careful comparison of all the sulphates, the proportions of which are well known, I deduced the weight of the atom of sulphuric acid to be 34; it now appears that the same weight is obtained synthetically, or without any reference to its combinations; the perfect agreement of these deductions, renders it beyond doubt that the weight is nearly approximated, and confirms the composition of the atom which has just been stated.

There are scarcely any chemical principles,

the proportions of which have been so diversely determined by experimentalists, as those of sulphuric acid : the following table will sufficiently prove the observation ; according to

Berthollet	72	sulphur	+ 28	oxygen.
Tromsdorf	70	———	+ 30	———
Lavoisier	69	———	+ 31	———
Chenevix	61.5	———	+ 38.5	———
Thenard	55.6	———	+ 44.4	———
Bucholz	42.5	———	+ 57.5	———
Richter	42	———	+ 58	———
Klaproth	42	———	+ 58	———

Chenevix's result would have been 44 sulphur + 56 oxygen, if he had adopted 33 per cent. acid in sulphate of barytes, which is now generally admitted. The method which he and the later experimentalists have taken, is to distil nitric acid from a given weight of sulphur, till the whole or some determined part of the sulphur is converted into sulphuric acid; the acid is then saturated with barytes, and the weight of the salt ascertained.

Notwithstanding the above theory of the formation of sulphuric acid was such as to convince me of its accuracy, I was desirous to see the manufacture of it on a large scale, and by the generous invitation of Mr. Watkins, of Darcy Lever, near Bolton, I had lately an opportunity of gratifying myself by the inspection of his large and well-conducted acid manufactory, near that place. When opening a small door of the leaden chambers, there issued a volume of red fumes into the air, which by their colour and smell, left no room to doubt of their being the fumes of nitrous acid. There was scarcely any smell of sulphurous acid. From the nitrous fumes, one would have been inclined to think that the chambers were filled with nitrous gas. I was particularly anxious to know the constitution of the air in the interior of the chambers, and Mr. Watkins was so obliging as to send me a number of phials of air taken from thence. Upon examination, the air was found to consist of 16 per cent. oxygen and 84 azote. There was no smell of sulphurous acid, and very little of nitrous acid, this last having been condensed in passing through the water. In fact, it seems that the nitrous acid fumes never make more, perhaps, than 1 per cent. upon the whole volume of air; nor can the oxygen be ever reduced much below 16 per cent, because the combustion would instantly cease. A constant dropping is observed from the roof of the chambers internally; these drops being collected, were found to be of the specific gravity 1.6; they had no sulphurous smell, but one slightly nitrous.

It is not very easy to suggest any plausible alteration in the management of a manufactory of this article.—Nitrous acid must be present; but whether it is best obtained by exposing nitre to the burning sulphur, or by throwing in the vapour of nitrous acid by direct distillation, may be worth enquiry. Loss of nitrous acid is unavoidable, partly by its escape into the air during the periods of ventilation, and partly by its condensation in the watery acid, on the floors of the chambers; a regular supply must, therefore, be provided; but if this exceed a certain quantity, it not only increases the expence, but is injurious to the sulphuric acid in some of its applications. There must, in all probability, be some figure of the chambers better than any other, in regard to their proportions as to length, breadth, and height; this, perhaps, can be determined only by experience. As water absorbs the nitrous acid with avidity, high chambers, and the combustion carried on at a distance from the water, must be circumstances favourable to economy in regard to nitre.

Sulphuric acid has a strong attraction for water; it even takes it from the atmosphere in the state of steam, with great avidity, and is therefore frequently used in chemistry for what is called *drying* the air. When mixed with water, sulphuric acid produces much heat, as has already been stated in the first part of this work.

When sulphuric acid is boiled upon sulphur, it has been said sulphurous acid is formed: I have not found this to be the case. But charcoal and phosphorus decompose the acid by heat; and the results are carbonic acid, phosphoric acid, and sulphurous acid.

Sulphuric acid combines with the alkalies and earths in general, forming with them salts denominated *sulphates*. On the metals this acid acts variously, according to its concentration; when diluted with 5 or 6 times its bulk of water, it acts violently on iron and zinc; great quantities of hydrogen gas are produced, which proceed from the decomposition of the water, and the oxygen of the water unites with the metal, to which the acid also joins itself, and a sulphate is thus formed. When the acid is concentrated, its action on metals is less violent; but by the assistance of heat, it oxidizes most of them, and gives off sulphurous acid.

As the sulphurie acid exists in various degrees of concentration, it becomes a matter of importance both to its manufacturer, and to those who use it largely, as the dyers and bleachers, to know the exact strength of it ; or in other words, to know how much water is combined with the pure acid in any specimen. This subject engaged the particular attention of Kirwan some years ago, and he has furnished us with a table of the strengths of sulphuric acid, of most densities. There are two things requisite to form an accurate table, the one is to ascertain the exact quantity of real acid in some specimen of a given specific gravity ; the other is to observe carefully the effects produced on the specific gravity of such acid, by diluting it with a given quantity of water. Mr. Kirwan has succeeded very well in the former, but has been peculiarly unfortunate in the latter. The errors of his table seem to have been known for the last 10 years to every one, except the editors of works on chemistry The following table exhibits the results of my own experience on this acid for several years.

Table of the quantity of real acid in 100 parts of liquid sulphuric acid, at the temperature 60°.

Atoms. Acid. Water	Acid per cent. by weight.	Acid per cent. by measure.	Specific gravity.	Boiling point.
1 + 0	100	unknown.	unknown.	unknown.
1 + 1	81	150	1.850	620°
	80	148	1.849	605°
	79	146	1.848	590°
	78	144	1.847	575°
	77	142	1.845	560°
	76	140	1.842	545°
	75	138	1.838	530°
	74	135	1.833	515°
	73	133	1.827	501°
	72	131	1.819	487°
	71	129	1.810	473°
	70	126	1.801	460°
	69	124	1.791	447°
1 + 2	68	121	1.780	435°
	67	118	1.769	422°
	66	116	1.757	410°
	65	113	1.744	400°
	64	111	1.730	391°
	63	108	1.715	382°
	62	105	1.699	374°
	61	103	1.684	367°
	60	100	1.670	360°
1 + 3	58.6	97	1.650	350°
	50	76	1.520	290°
	40	56	1.408	260°
1 + 10	30	39	1.30+	240°
1 + 17	20	24	1.200	224°
1 + 38	10	11	1.10—	218°

Remarks on the preceding Table.

1. The acid of 81 per cent. is constituted of 1 atom of acid and 1 of water. It is the strongest possible acid that can be obtained by boiling the liquid acid; because at that strength

the acid and water distil together, in the same way as nitric acid of 1.42 sp. gravity, or muriatic of 1.094. It is a mistaken notion, that the common sulphuric acid of commerce is of the maximum strength, though it is of the maximum density nearly. The fact is, acid nearly of the maximum strength varies very little in its specific gravity, by the addition or subtraction of a small quantity of water. Here is Kirwan's principal error. Acids of the strength of 81 and 80, do not differ more than 1 in the third place of decimals; whereas, according to his table, the difference is 14 times as great. The acid of commerce varies from 75 to 80 per cent. of acid, or about 7 per cent. in value, in the different specimens I have had occasion to examine. This variation only changes the second figure in decimals an unit; though, according to Kirwan's table, the change is 7 times as much. The specific gravity ought not to be the criterion of strength in acids above 70 per cent.; the temperature at which they boil is a much better criterion, because it admits of a range of 12 or 15° for 1 per cent. of acid. Or the strength may be found by determining what quantity of water must be added to reduce the acid to some known strength, as that of the glacial acid, of 1.78 sp. gravity.

2. There is nothing further striking in the table till we come to the acid, which is constituted of 1 atom to 2 of water; this acid possesses the remarkable property of congealing in a temperature at or above 32°, and of remaining congealed in any temperature below 46°; its specific gravity is 1.78, as Keir found it, (Philos. Trans. 1787), and it contains 68 per cent. of real acid, both by theory and experiment; it is determined by theory thus: — one atom of sulphuric acid weighs 34, and 2 of water 16, together making 50; hence, if 50 : 34 : : 100 : 68; it is found experimentally thus: let 100 grain measures of glacial sulphuric acid be saturated with carbonate of potash, and the sulphate of potash be obtained; it will weigh, after being heated to a moderate red, nearly 270 grains, of which 121 will be acid, and 149 alkali, according to the analyses of Kirwan and Wenzel. If the liquid acid be of greater or less specific gravity, so as to contain even 1 per cent. more or less real acid, then it cannot be frozen in a temperature above 32°, but may in a temperature a little below 32°. If the liquid acid contain 3 per cent. more or less than the glacial, it cannot be frozen without the cold produced by a mixture of snow and salt; and that is insufficient, if it deviate more than 3

per cent. from the glacial, as Mr. Keir determined. I find the frozen acid to be of the specific gravity 1.88 nearly. It seems probable that the difficulty of freezing would increase in both sides, till the acids of 1 and 1 above, and 1 and 3 below.

3. The acids below 30 per cent. may, without any material error, have their strength estimated by the first and second figures of decimals in the column of sp. gravity; thus acid of 15 per cent. strength, will have the specific gravity 1.15, &c.

SECTION 5.

OXYGEN WITH PHOSPHORUS.

There are only two compounds of oxygen and phosphorus yet known: they both have the characters of acids; the one is denominated phosphorous acid, the other phosphoric acid. It is extremely probable that the former, though recognised as an acid, is yet in the lowest degree of oxidation, and may therefore with equal propriety be called *phosphorous oxide*, *phosphoric oxide*, or, after the manner of metals, *oxide of phosphorus*. We shall, however, adopt the common name.

1. *Phosphorous Acid.*

When phosphorus is exposed for some days to the atmosphere, it gradually acquires oxygen, and is converted into an acid liquid. This process may be effected by putting small pieces of phosphorus on the sloping sides of a glass funnel, and suffering the liquid to drop into a phial as it is formed. The liquid, called phosphorous acid, is viscid, tastes sour, and is capable of being diluted with water to any amount. It has the usual effect of acids on the test colours. When heated, water is evaporated, and afterwards phosphuretted hydrogen gas; finally, there remains phosphoric acid in the vessel. It should seem from this, that heat gives the oxygen of one part of the phosphorous acid to another, by which the latter is changed into phosphoric acid, and the phosphorus of the former is liberated; but at that degree of heat the liberated phosphorus acts on the water; one part of it takes the oxygen to form more phosphorous acid, and the other takes the hydrogen to form phosphuretted hydrogen; and thus the process is carried on till all the phosphorus is in the state of phosphoric acid, or phosphuretted hydrogen. It is probable, that in this way the phosphorus is divided, so that two thirds of it are united to oxygen, and one third to hydrogen ; but this has not been ascertained by direct experiment.

Phosphorous acid acts upon several metals, oxidizing them by the decomposition of water, and at the same time giving out phosphuretted hydrogen ; the resulting metallic salts are, it is supposed, phosphates, the redundant phosphorus being carried off by the hydrogen. This acid combines with the alkalies, earths, and metallic oxides, and forms with them a class of salts called *phosphites*.

When nitric acid is put to phosphorous acid, and heat applied, the nitric acid is decomposed, half of its oxygen unites to the phosphorous acid, and converts it into phosphoric acid, and the rest of the nitric acid escapes in the form of nitrous gas.

The proportion of the two elements constituting phosphorous acid has not hitherto been ascertained ; I am inclined to believe, from the experiments and observations about to be related concerning phosphoric acid, that phosphorous acid is composed of 1 atom of phosphorus, weighing 9 nearly, and 1 of oxygen, weighing 7 ; the compound weighing 16. If this be the case, it may appear singular that none of the other elements exhibit acid properties when combined with 1 atom of oxygen ; but it should be observed, that the phosphoric oxide is in a liquid form, and disposed to separate into phosphorus and phosphoric acid, circumstances that do not combine in regard to the other oxides. In fact, phospherous acid may be considered as phosphoric acid holding phosphorus in solution, rather than as a distinct acid.

2. *Phosphoric Acid.*

Though some of the compounds of phosphoric acid, and the earths and alkalies, are common enough, yet this acid, in a pure state, is rarely obtained in any considerable quantity, requiring a process both tedious and expensive. There are three methods by which phosphoric acid may be formed : 1. If a small portion of phosphorus, namely, from 5 to 20 grains, be ignited, and immediately covered with a large bell glass, over water, the phosphorus burns with great brilliancy, and soon fills the vessel with white fumes ; in a short time, the combustion ceases ; after which the fumes gradually subside, or adhere to the side of the glass in the form of dew ; these white fumes are pure phosphoric acid. 2. If a small piece of phosphorus be dropped into heated nitric acid in a phial or gas bottle, a brisk effervescence ensues, occasioned by the escape of nitrous gas, and the phosphorus gradually disappears, being converted into phosphoric acid, and mixed with the remaining nitric acid; another small piece may then be dropped into the liquid, and so on in succession till the nitric acid is almost wholly decomposed; the remaining liquid may then be gradually increased in temperature, to drive off all the nitric acid; what is left is a liquid consisting of phosphoric acid and water; by increasing the heat to a moderate red, the water is driven off, and liquid phosphoric acid remains, which on cooling becomes like glass. 3. If phosphorous acid be prepared by the slow combustion of phosphorus, as mentioned above, and then a portion of nitric acid added to the liquid, and heat be applied, the nitric acid gives part of its oxygen to the phosphorous acid, and nitrous gas escapes. What remains, when heated, is pure phosphoric acid.

Of these three processes, the first may be recommended when the object is to find the proportion of the elements of the acid; but the second and third, when the object is to procure a quantity of acid for the purposes of investigation Of these the third is preferable in an economical point of view, because it requires only half as much nitric acid. By calculation, I find that 20 grains of phosphorus will require 200 grains of nitric acid of 1.35, by the second process, but only 100 grains by the third; but a small excess should always be allowed for loss by evaporation, &c.

Phosphoric acid, in the state of glass, is deliquescent when exposed to the air; it becomes oily, and may be diluted with any quantity of water. This acid is not so corrosive as some others; but it has the other acid properties of a sour taste, of reddening vegetable blues, and of combining with the alkalies, earths, and metallic oxides, to form salts, which are called *phosphates*. It has the power of oxidizing certain metals, by decomposing water in the manner of sulphuric acid; the oxygen of the water unites to the metal, and the hydrogen is liberated in the state of gas. Charcoal decomposes this acid, as well as the phosphorous, in a red heat; hence the process for obtaining phosphorus form superphosphate of lime.

Nothing very certain has been determined respecting the relation of the strength of this acid to the specific gravity of the liquid solution. Some experience I have had, makes me think the following table will be found nearly correct : at all events, it may have its use till a better can be formed.

Table of the quantity of real acid in 100 parts of liquid phosphoric acid.

Acid per cent. by weight.	Acid per cent. by measure.	Specific gravity.
50	92.5	1.85
40	64.	1.60
30	41.7	1.39
20	24.6	1.23
10	11.	1.10

Lavoisier ascertained the relative weights of phosphorus and oxygen in phosphoric acid to be 40 to 60 nearly : this was effected by burning phosphorus in oxygenous gas. This important fact has been since corroborated by the experience of others. I find a near approximation to this result by burning phosphorus in atmospheric air. In a bell glass, containing 400 cubic inches of air, 5 grains of phosphorus were repeatedly burnt over water ; the combustion at first was very vivid, but towards the conclusion it was languid ; there was a residuum of moist, half burned phosphorus in the cup, usually about 1 grain : the glass had a flaccid bladder adapted to it to receive the rarefied air, so as to suffer none to escape. The air at first contained $20\frac{1}{2}$ per cent. oxygen; but after the combustion, it contained only 16 or $16\frac{1}{2}$ per cent., the temperature being about 40° at the time. Whence, by calculation, it appears that in these instances 4 grains of phosphorus may be concluded to have united to 6 grains of oxygen. The data, indeed, would give a rather less proportion of oxygen; but it is probable that some phosphorous acid is formed near the conclusion of the combustion.

With respect to the constitution of the phosphoric acid atom, there can be but two opinions entertained. Either it must be 1 atom of phosphorus with 2 atoms of oxygen, or with 3 of oxygen. According to the former opinion, the phosphoric atom will weigh 9, and the phosphoric acid atom 23; according to the latter opinion, the phosphoric atom will weigh 14, and the acid atom 35. We might appeal to the phosphates to determine the weight of the acid; but this class of salts has not been analyzed with sufficient precision. Fortunately, there is another compound of phosphorus which is subservient to our purpose; namely, phosphuretted hydrogen. As the properties of this gas will be treated of in the proper place, we shall only observe here that the gas is a compound of phosphorus and hydrogen; that it contains just its bulk of hydrogen; that its specific gravity is about 10 times that of hydrogen; and that when fired in Volta's eudiometer along with oxygen, it is converted into water and phosphoric acid, requiring 150 per cent. in volume of oxygen for its complete combustion; but is, notwithstanding, burnt so far as to lose its elasticity with 100 measures of oxygen. These facts leave no doubt that the atom of phosphorus weighs 9; that the atom of phosphoric acid weighs 23, being a compound of 1 with 2 of oxygen; that the atom of phosphorous acid is 1 with 1 of oxygen, weighing 16, and that phosphorous acid and water are formed when equal volumes of phosphuretted hydrogen and oxygen are exploded together.

SECTION 6.

HYDROGEN WITH AZOTE.

Only one compound of hydrogen and azote has yet been discovered: it has been long known to chemists as an important element, and under various names, according to the state in which it was exhibited, or to the article from which it was derived; namely, *vo-*

latile alkali, *hartshorn*, *spirit of sal ammoniac*, &c. but authors at present generally distinguish it by the name of *ammonia*. Its nature and properties we shall now describe.

Ammonia.

In order to procure ammonia, let one ounce of powdered sal ammoniac be well mixed with two ounces of hydrate of lime (dry slaked lime), and the mixture be put into a gas bottle; apply the heat of a lamp or candle, and a gas comes over, which must be received in jars over dry mercury. It is *ammoniacal gas*, or ammonia in a pure state.

This gas is unfit for respiration, and for supporting combustion; it has an extremely pungent smell, but when diluted with common air, it forms an useful and well-known stimulant to prevent fainting. The specific gravity of this gas has been found nearly the same by various authors, which is the more remarkable, as the experiment is attended with some difficulties that do not occur in many other cases. According to Davy, 100 cubic inches of it weigh 18 grains; according to Kirwan, 18.2 grains; Allen and Pepys, 18.7; and Biot, 19.6; the mean of these, 18.6 grains, may be considered as a near approximation at the temperature 60° and pressure 30 inches of mercury: hence the specific gravity is .6, the weight of atmospheric air being one.

Ammoniacal gas sent into water, is condensed almost with the same rapidity as steam; in this respect it corresponds with fluoric and muriatic acid gases. The compound of water and ammonia forms the common liquid ammonia sold by the name of *spirit of sal ammoniac;* this is the form in which ammonia is the most frequently used. It is of great importance to ascertain the quantity of gaseous or real ammonia in given solutions of ammonia in water. This subject has been greatly neglected; a very good attempt was made about 10 years ago by Mr. Davy, to ascertain the quantity of ammonia in watery solutions, of different specific gravities; the result was a table, which may be considered an excellent first approximation; but it is to be regretted that so important an enquiry should not have attracted attention since. I have instituted a few experiments on this head, the results of which will no doubt be acceptable.

A phial, capable of holding 1400 grains of water, was partly filled with mercury, and the rest with 200 grains of water, and inverted in mercury; into this 6000 grain measures of ammoniacal gas were transferred; the liquid had not diminished sensibly in specific gravity; it required $24\frac{1}{2}$ grain measures of muriatic acid, 1.155, to saturate the water; by evaporating in a heat below boiling water, 12 grains of dry muriate of ammonia were obtained. Now, supposing 1400 measures of gas equal to 1 grain in weight, there would be found in the salt 5.7 grains of muriatic acid, 4.3 grains of ammonia, and 2 grains of water. I found this method of proceeding not to be relied upon; for, though the mercury had recently been dried in an oven in the temperature 240°, yet the ammoniacal gas could not be transferred from one graduated tube to another, without a loss of 10 or 15 per cent.; I had reason to conclude, then, that the ammonia in the above salt was overrated. In order to avoid this source of error, I adopted the method first used by Dr. Priestley, of putting muriatic acid gas to the alkaline in the graduated tube; but here was still an objection, as the muriatic acid gas must be measured previously to the transfer, and it is equally absorbable by water with alkaline gas. However, I found, as Dr. Priestley had done before, that equal measures of the two gases as nearly as possible saturated each other. For, when a measure of acid gas was put to one of alkaline, there was a small residuum of alkaline gas ; and when the alkaline was transferred to the acid, there was a small residuum of acid gas. Having before concluded (page 287) that muriatic acid gas was of the specific gravity 1.61, I might have adopted the ratio of acid and alkali in muriate of ammonia to be 1.61 to .6 ; and hence have inferred the quantity and volume of ammonia in a given solution, from the quantity of muriatic acid solution requisite to saturate it. But there was one important circumstance against this ; the atom of muriatic acid I knew weighed 22, and the ratio of 1.61 to .6, is the same as 22 to 8.2 nearly ; hence, the weight of an atom of ammonia must have been 8.2 or 4.1, which I was aware was inconsistent with the previous determinations concerning azote and hydrogen. Observing in the 2d vol. of of the *Memoires d'Arcueil,* that Biot and Gay Lussac find the specific gravity of muriatic acid gas to be so low as 1.278, and understanding from conversation with Mr. Davy, that he also had found the specific gravity of the gas to be considerably less than I had concluded, I was induced to repeat the experiment of weighing it, taking every care to avoid the introduction of liquid solution. I sent a stream of acid gas, derived from common salt and concentrated sulphuric acid, through an intermediate vessel, into a dry flask of common air, loosely corked, till it had expelled $\frac{3}{4}$ths of the air, as appeared afterwards; the inside of the glass had a very slight opacity on its surface; it had gained $1\frac{1}{10}$ grain in weight; it was then uncorked and its mouth plunged into water, when $\frac{3}{4}$ths of the flask were in a few moments occupied by the water. Other trials gave similar results. The flask held 6 grains of common air. Whence I derive the specific gravity of muriatic acid gas to be 1.23, and am induced to apprehend that this is rather more than less than the truth. The weights of equal volumes of muriatic acid gas and ammoniacal gas will then be as 1.23 to .6; or as 22 to 11, nearly; and if we assume that 11 measures of acid gas are sufficient for 12 of alkaline, which is not unlikely from experience; then we shall have 22 parts acid to 12 of ammonia for the constitution of muriate of ammonia (exclusive of water), which will make the theory and experience harmonize; according to this view, muriate of ammonia must consist of 1 atom of muriatic acid and 2 of ammonia, each atom of ammonia being a compound of 1 atom of azote and 1 of hydrogen. However this may be, I find that 22 parts of real muriatic acid, 38 of nitric, and 34 of sulphuric, as determined by the respective foregoing tables, will saturate equal portions of any ammoniacal solution ; these, then, may be considered as tests of the quantity of real ammonia in different solutions; and if the ratio of 22 to 12, above adopted, be incorrect, it cannot be greatly so ; and the error will be general, being so much per cent. upon any table of ammoniacal solutions. The test acids I prefer for use, are such as contain half the quantities of acid above stated in 100 grain measures. Thus, 100 grain measures of muriatic acid, sp. gravity 1.074, contain 11 grains of real acid ; 100 measures of nitric acid, 1.141, contain 19 grains; and 100 measures of sulphuric acid, 1.135, contain 17 grains of real acid. Now, 100 measures of ammoniacal solution of .97 sp. gravity, are just sufficient to saturate these. Whence I adopt that solution as test ammonia, and conclude that 100 grain measures of it contain 6 grains of real ammonia.

It will be perceived, then, that the accuracy of the ensuing table depends upon several points: namely, whether 100 measures of muriatic acid of 1.074, really contain 11 grains of acid ; whether the specific gravities of muriatic acid gas, and ammoniacal gas, are really 1.23 and .6, or in that ratio ; and whether 11 measures of acid gas saturate 12 measures of ammoniacal gas. I believe the errors in any of these particulars to be very small, and probably they may be such as partly to correct each other.

I find, after Mr. Davy, that a measure of water being put to a measure of ammoniacal solution, the two occupy two measures, without any sensible condensation; consequently, if the quantity of ammonia in a measure of any given specific gravity, as .90, be determined; then the quantity in a measure of .95, will be just half as much: Hence, a table is easily constructed for measures, and one for weights is derivable without much calculation.

Table of the quantities of real or gaseous ammonia in solutions of different specific gravities.

Specific gravity.	Grains of ammonia in 100 water grain measures of liquid.	Grains of ammonia in 100 grains of liquid.	Boiling point of the liquid old scale.	Volume of gas condensed in a given volume of liquid.
.85	30	35.3	26°	494
.86	28	32.6	38°	456
.87	26	29.9	50°	419
.88	24	27.3	62°	382
.89	22	24.7	74°	346
.90	20	22.2	86°	311
.91	18	19.8	98°	277
.92	16	17.4	110°	244
.93	14	15.1	122°	211
.94	12	12.8	134°	180
.95	10	10.5	146°	147
.96	8	8.3	158°	116
.97	6	6.2	173°	87
.98	4	4.1	187°	57
.99	2	2	196°	28

On the above table, it may be proper to remark, that I have not had large quantities of ammoniacal solution lower than .94, so as to find their specific gravities experimentally; but have had small quantities to the amount of 10 or 20 grains of the several solutions from 26 to 12 per cent.; I have no reason to suspect any material deviation from the law of descent observed in the specific gravity down to 12 per cent., when we go below that number; at all events, it cannot be great down to .85, and it is not of much importance, because solutions of that strength are never obtained in the large way.—The second column, exhibiting the grains of ammonia in 100 measures of the solution, is more convenient for practice than the third, which gives the weight in 100 grains of solution. The fourth column, which shews the temperature at which the several solutions boil, will be found highly interesting. The ebullition of a liquid is well known to take place, when the steam or vapour from it is of the same force as the atmospheric pressure. In solutions down to 12 per cent., the experiments were performed by inserting a thermometer into a phial containing the solution, and plunging the phial into hot water till the liquid boiled; but in the higher solutions a small portion, as 20 grains, was thrown up a tube filled with mercury; the tube was then put into a phial of mercury, and the whole plunged into warm water; the temperature was then ascertained requisite to bring the mercury in the tube to the level of that in the phial. The fifth column is calculated from the second, supposing the specific gravity of ammoniacal gas = .6.

It may be observed, that the above table gives the quantity of ammonia in different solutions, from 15 to 20 per cent. less than Mr. Davy's table; also, that the common ammoniacal solutions of the shops usually contain from 6 to 12 per cent. of ammonia.

Before we can estimate the value of the fourth and fifth columns of the table, we must ascertain the force of vapour from ammoniacal solutions at different temperatures. If it be found in some one instance, we may by analogy infer the results in others. As the steam from water varies in force in geometrical progression to equal increments of temperature, it might be expected that the steam or gas from liquid ammonia should do the same; but as the liquid is a compound, the simple law of the force of aqueous steam does not obtain. It appears, however, from the following results, that a near approximation to this law is observed. Into a syphon barometer I threw a quantity of 946 liquid ammonia, which was by agitation, &c. transferred to the vacuum over the mercury. The vacuum was then immersed successively in water of different temperatures, and the force of the gas observed as under.

Temperature.			
old scale.	new scale.	differences.	Force of ammoniacal steam from liquid .946.
140°	151°		30 inch.
		36°	
103°	115°		15
		31°	
74°	84°		7.5
		29°	
50°	55°		3.75

Hence it seems, that the intervals of temperature required to double the force of ammoniacal steam, increase in ascending. I had no doubt but this sort of steam or gas, would mix with common air, without having its elasticity affected, like as other steams do ; but I ascertained the fact by experiment : Thus I mixed a given volume of air with steam of 15 inches force, and found that the air was doubled in bulk.

These facts are curious and important. They shew that ammonia is not retained in water without external force, and that the pressure of no elastic fluid avails but that of ammoniacal gas itself; thus establishing the truth of the general law which I have so much insisted on, that *no elastic fluid is a sufficient barrier against the passage of another elastic fluid.*

We may now see upon what causes the saturation of water with ammonia depends. They are two; the *temperature* of the liquid; and the *pressure* of the incumbent ammoniacal gas, exclusive of the air intermixed with it. For instance, if the temperature be given, 50° (old scale); then the strongest possible solution, under atmospheric pressure, will be such, that 100 measures will have the specific gravity 87, and contain 26 grains of ammonia, or 419 times the volume of gas. But if, in saturating the water by sending up gas, there be common air, so as to make $\frac{7}{8}$ths of the incumbent gas, then the solution cannot be made stronger than .946, of which 100 measures contain 11 grains of ammonia, or 162 times the volume of gas. I have obtained a saturated solution containing 26 per cent. ammonia, with $\frac{1}{12}$th common air in the incumbent gas; and at the same temperature, another saturated solution, containing only 17 per cent. ammonia, with $\frac{3}{4}$ths common air in the incumbent gas.

With respect to the constitution of ammonia, Priestley, Scheele, and Bergman pointed out the two elements into which it is decomposed. Berthollet first settled the proportions of the elements, and the quantity of each obtained from a given volume of ammoniacal gas. It is highly to his credit too, that subsequent repetitions of his experiments, under the improved state of knowledge, have scarcely amended his results. Priestley resolved 1 measure of ammoniacal gas, by eleetricity, into 3 measures of gas not absorbable by water; but his ammonia could not have been dry. Berthollet resolved 17 measures into 33 in the same way: this result has since been corroborated by various authors. He also found that the gas so produced, was a mixture of 121 parts of azote by weight, with 29 of hydrogen; or 4.2 azote with 1 of hydrogen.

In 1800, Mr. Davy published his researches, in which were given several interesting results on ammonia, Mr. Davy decomposed ammonia, by sending the gas through a red hot porcelain tube; after the common air was expelled, the collected gas was found free from oxygen. To 140 measures of this gas were added 120 of oxygen; the mixture being exploded by electricity, 110 measures of gas were left; and of course 150 were cenverted into water; of this 100 measures must have been hydrogen. Whence, 140 measures of the gas from decomposed ammonia, contained 100 hydrogen and 40 azote; or 100 measures contained 71.4 hydrogen and 28.6 azote. This conclusion was so nearly agreeing with the previous determination of Berthollet, that both have justly been held up as specimens of the accuracy of modern chemical analysis.

In 1808, Mr. Davy published his celebrated discoveries relating to the decomposition of the fixed alkalies. Having ascertained that these contained oxygen, he was led by analogy to suspect the same element in ammonia. Several experiments were made, which seemed to countenance this idea; but these could not be considered conclusive, as long as it was admitted that no oxygen appeared in the decomposition of ammonia by electricity, and yet that the weight of the azote and hydrogen were together equal to that of the ammonia decomposed. Mr. Davy re-examined the specific gravity of ammoniacal gas, the quantity of gases evolved by the decomposition of a given volume of it, and the ratio of azote to hydrogen in the same. The result was, that the gases obtained amounted only to $\frac{10}{11}$ths of the weight of the ammonia; the remaining $\frac{1}{11}$th Mr. Davy thought must be oxygen, which, uniting to hydrogen, formed a portion of water. The way in which this $\frac{1}{11}$th was saved, was principally by diminishing the absolute quantity of gases derived from a given volume of ammonia, but partly by finding less azote and more hydrogen than had been before estimated. Thus, 100 measures of ammoniacal gas produced only 180 measures of mixed gas, though commonly estimated at 200; and this gas was found to consist of 26 azote and 74 hydrogen per cent.

These conclusions, so different from what had been long adopted, and depending upon experiments of some delicacy, were not likely to be received without a more general scrutiny. Dr. Henry in England, and A. B. Berthollet in France, seem both to have renewed the investigation into the component parts of ammonia with great care and assiduity. Dr. Henry's object was to determine whether any oxygen, water, or any other compound containing oxygen, could be detected in the analysis of ammonia; this enquiry included the two others; namely, the quantity of gases obtained from a given volume of ammoniacal gas, and the proportion of azote to hydrogen in the same. The results were, that neither oxygen nor water could be found; that for the most part the bulk of ammonia was doubled by decomposition, even when the gas was previously dried with extreme care; and that the ratio of azote to hydrogen in the mixture, from an average of six careful experiments, was $27\frac{1}{4}$ to $72\frac{3}{4}$. In this last decision, Dr. Henry was so fortunate as to discover a more easy and expeditious mode of analysis than had been known before; he found that ammoniacal gas mixed with a due proportion of oxygen, of nitrous oxide, or even of nitrous gas, would explode by an electric spark. He found an under proportion of oxygen gas to answer best (about 6 measures of oxygen to 10 of ammonia): the explosion produced a complete decomposition of the ammonia, and a partial combustion of the hydrogen; after which more oxygen was put to the residuum, and the remainder of the hydrogen consumed. From one experiment, in which 100 measures of ammonia were decomposed in a tube of which the mercury had been previously boiled, Dr. Henry only obtained 181 measures of gas; and he seems to think that this experiment may be the most correct in regard to that object. (Philos. Trans. 1809).

In the Memoires d'Arcueil, tom. 2, M. A. B. Berthollet has a paper on the analysis of ammonia. He alludes to the experiments of Berthollet in the memoirs of the academy,

1785; in which the ratio of 27.5 azote to 72.5 hydrogen, was found in the decomposed ammonia, allowing 196 hydrogen for 100 oxygen. He reports several experiments and observations relative to the oxidation and deoxidation of iron in ammoniacal gas. He then proceeds to prove, that the weight of azote and hydrogen produced in the decomposition of ammonia, is equal to the weight of the ammonia itself. Biot and Arago determine the specific gravities of azote, hydrogen, and ammonia, to be .969, .073, and .597 respectively, which A. B. Berthollet adopts. He finds that 100 measures of ammonia produce 205 of permanent gas; which, by analysis, gives 24.5 azote and 75.5 hydrogen per cent. Like Dr. Henry, A. B. Berthollet decomposed ammonia by exploding it with oxygen gas; but unfortunately he used an *excess* of oxygen, and then determined the residuary oxygen by the addition of hydrogen: he was aware, however, that part of the azote was thus converted into nitric acid. Upon collecting the results, he makes it appear, that the gases produced by the decomposition of ammonia are, as nearly as possible, equal to the weight of the ammonia.

Though the experiments of these two authors may be deemed satisfactory, with regard to the non-existence of oxygen in ammonia, they would have been more so if they had accorded in the quantity of gas derived from a given volume of ammonia, and in the ratio of azote to hydrogen. Having made some experiments myself on these heads, I may be allowed to give my opinion as to the causes of these differences.—I am persuaded, with Mr. Davy, that ammonia is not doubled by decomposition, when due care is taken to prevent any liquid from adhering to the tube or mercury; but at the same time am inclined to believe, from experience, that 100 measures of ammonia will give not less than 185 or 190 measures of gas by dscomposition: I took a tube and filled it with dried mercury; then transferred a portion of gas into it, and by pushing a glass rod up the tube several times, displaced the mercury in the tube, so that no liquid ammonia could exist in the renovated mercury. This gas, being decomposed by electricity, produced after the rate of 187 for 100. With respect to the ratio of azote to hydrogen, I am convinced it is to be obtained only by decomposing the ammonia previously to the combustion of the hydrogen, and this may be done either by electricity or by heat; in these cases, ammonia will be resolved into 28 measures of azotic gas, and 72 measures of hydrogen gas, in the hundred. I have repeatedly obtained it so by electricity, the results never deviating farther than from 27 to 29 of azote. This agrees sufficiently with Berthollet's original analysis by electricity, and with Davy's analysis by heat in 1800; both of them made without any theoretic views as to quantity, which cannot be said of any of the subsequent investigations on this subject.

We are now to see how far these results will agree with the specific gravity of ammoniacal gas: that is, whether the weights of the two gases are equal to the weight of the ammonia decomposed.

			Grains.
	100 measures of ammonia,	which × sp. gr. .6	gives 60
become	185 measures of mixed gas,		——
namely,	51.8 azote, — — —	which × sp. gr. .967	gives 50.09
and	133.2 hydrogen, — —	which × sp gr. .08	gives 10.65
			——
			60.74

The excess of $\frac{3}{4}$ths of a grain upon 60, is too small to merit notice, and may arise from an inaccuracy in any of the data, which, if corrected, could have no material influence on the conclusions.

I shall now make a few observations on the other methods of analyzing ammonia. Dr. Henry's methods of burning ammonia in Volta's eudiometer along with oxygenous gas, nitrous gas, and nitrous oxide, unite elegance with expedition, and when well understood, cannot but be valuable. It appears to me, however, both from experience and analogy, that a compound combustible, such as ammonia, is never decomposed and one of its elements burnt, to the entire exclusion of the other. Numerous instances may be found in the compounds of charcoal and hydrogen, of phosphorus and hydrogen, &c. where one of the elements seizes the oxygen with more rapidity than the other ; but some portion of the other is always burnt. Even when the combustible gases are only mixed together, and not combined, we do not find that one of them precludes the other from taking a share of the oxygen till it is saturated. Thus, in a mixture of carbonic oxide with hydrogen, with a deficiency of oxygen, part of both is burnt by an electric spark. Dr. Henry has, indeed, noticed that ammonia fired with excess of oxygen, gives nitric acid as well as water. I have reason to believe this is the case in some degree, in whatever proportion they are fired. I have seldom obtained so much as 27 per cent. of azote by the combustion of ammonia with oxygen (the hydrogen being estimated by doubling the oxygen spent), and in no instance 28 : but it will be manifest that all the oxygen is not consumed in burning the hydrogen, if we note the ammoniacal gas expended, and allow only 66 or 67 per cent. oxygen for the hydrogen; there will generally be found a greater expence of oxygen, which must have gone to form nitric acid. The combustion of ammonia with nitrous gas usually gives from 25 to 27 per cent. of azote, allowing the constitution of nitrous gas to be what is stated at page 331. Upon the whole, I found nitrous oxide to approximate nearest to the truth. When 100 measures of ammonia are exploded with 120 of nitrous oxide, the gases resulting are azote with a very small portion of hydrogen; if to this a little hydrogen be added, and then an excess of oxygen, another explosion will determine the residuary hydrogen; which being deducted, there remain about 172 azote, 120 of which come from the nitrous oxide, and 52 from the ammonia, which gives after the rate of 28 azote per cent. on the evolved gases.—When the decomposition of ammonia is attempted by oxymuriatic acid gas, a graduated tube is filled with the gas, and plunged into liquid ammonia; in this way, if we reckon a measure of the acid gas to a measure of hydrogen, we shall find the azote evolved and left in the tube, amount to 23 or 24 per cent upon both gases. It is to be presumed then, that oxymuriatic acid, like oxygen, consumes part of both the elements of ammonia.

By comparing the weight of azote with that of hydrogen in the above table, we find them as 4.7 to 1 nearly. This evidently marks the constitution of ammonia to be that of 1 atom of each of the elements combined. But we have before determined the element of azote to weigh 5.1, when treating of the compounds of azote and oxygen. This difference is probably to be ascribed to our having over-rated the specific gravity of nitrous gas, and perhaps nitrous oxide: In the *Memoires d Arcueil*, I observe Berard finds the specific gravity of nitrous gas to be 1.04, instead of 1.10, which last I have made my calculations from; if the former should prove true, it will reduce my valuation of azote in nitric acid nearly to 4.7; I have not had an opportunity of ascertaining the specific gravity of nitrous gas; but am inclined to believe that 1.10 may be too high. Berthollet finds nitrous oxide to be 1.36, instead of 1.61; I much suspect the former is too low.

Upon the whole, we may conclude that an atom of ammonia is constituted of 1 atom of hydrogen and 1 of azote, and weighs nearly 6 The diameter of the elastic particle is .909, that of hydrogen being 1. Or, 300 measures of ammoniacal gas contain as many atoms as 400 measures of hydrogen, or as 200 of oxygen.

SECTION 7.

HYDROGEN WITH CARBONE.

There are two combinations of hydrogen with carbone, now well known, easily distinguishable from each other and from all other combinations. They are both elastic fluids; one of them, called olefiant gas, is a compound of 1 atom of hydrogen and 1 of carbone; the other, which I call carburetted hydrogen, is a compound of 2 atoms of hydrogen and 1 of carbone, as will be manifest from what follows.

1. *Olefiant Gas.*

The gas denominated *olefiant*, was discovered and examined by the Dutch chemists, Bondt, Dieman, &c. and a memoir on the subject was published in the 45th vol. of the Journal de Physique, 1794.

Olefiant gas may be procured by mixing 2 measures of sulphuric acid with 1 measure of alcohol; this mixture in a gas bottle must be heated to about 300° by a lamp, when the liquid exhibits the appearance of ebullition, and the gas comes over: it should be passed through water, to absorb any sulphurous acid which may be generated.

This gas is unfit for respiration, and extinguishes flame, but it is highly combustible: its specific gravity, according to the Dutch chemists, is .905; according to Dr. Henry, 967. Perhaps .95 is about the truth. Water absorbs $\frac{1}{8}$th of its bulk of this gas; or the atoms of gas in the water are just twice the distance they are without; and it may be expelled again by the other gases. This property (of being absorbed by 8 times its bulk of water) occurred to me in 1804, in a course of experiments on the absorption of gases by water. It is peculiar to this gas, and consequently distinguishes it from all others. When olefiant gas is mixed with oxymuriatic acid gas, a diminution takes place, like as when oxygen and nitrous gas are mixed; but the result is an *oil*, which swims on the surface of the water. Hence the Dutch chemists gave this gas the name of *olefiant*. For this purpose, they found 3 measures of olefiant gas required

4 measures of the acid gas; but Dr. Henry finds 5 of olefiant require 6 of the acid. The difference is not great, considering the difficulty of the experiment. As neither of these results will agree with the other known properties of these two gases, I suspected that both would be found in some degree incorrect; which proved to be the case from the following experiments. Having taken two similar tubes graduated, containing each about 170 grains of water, I filled them, one immediately after the other, from a bottle generating oxymuriatic acid copiously; into one of these, 200 measures of olefiant gas were slowly transferred; after standing some time, the residuary gas was transferred and noted; then the other tube with acid gas was taken, the gas passed 5 or 6 times through water, till no further diminution was observed, and the residuary gas was noted and allowed for impurity in the first tube. By this procedure no acid gas was lost, and an excess of olefiant gas being used, the purity of this last did not enter into the calculation. In one trial, 165 measures of oxymuriatic acid gas condensed 168 of olefiant gas; in another, 165 took 167. From these, I conclude that oxymuriatic acid requires a very little more than its bulk of olefiant gas to be saturated: perhaps 100 of the former may take

102 measures of the latter ; but if we reckon equal volumes, the error cannot in general be material.

Olefiant gas burns with a dense, white flame. It explodes with uncommon violence when mixed with oxygen and electrified ; the products resulting are various, according to the circumstances. When completely saturated with oxygen, the results are, according to

			carb. acid.
Berthollet,	100 measures take	280 oxygen, produce	180
Dr. Henry,	100 — —	284 — —	179

The rest of the produce is water. These results, agreeing so well with each other, are the more plausible ; but I can add that my own experience corroborates them, particularly in regard to oxygen: My results have always given less than 300, but more than 270; the acid, I apprehend, should be about 185 or 190 : unless a great excess of oxygen be used, the charcoal is partly thrown down, and it makes the gas turbid after the explosion ; the result in this case affords less carbonic acid than is due.

When olefiant gas alone is subjected to continued electricity, either over mercury or water, the result is hydrogen gas, and a quantity of charcoal is deposited. A very careful experiment of this kind was made by Dr. Henry and myself, in which 42 measures of pure olefiant gas were electrified till they became 82 ; these were exploded with oxygen, and found to consist of 78 hydrogen, and 2 olefiant gas. Here 40 olefiant became 78 hydrogen, or very near double. The charcoal was thrown down. According to this, 100 measures of olefiant gas will contain 195 of hydrogen ; which require 98 oxygen for their combustion ; now as the charcoal must take the rest, or nearly 196 measures, it follows that in the combustion of olefiant gas, 2 parts of the oxygen are spent upon the charcoal, and 1 part upon the hydrogen. Hence we obtain this conclusion, that an atom of olefiant gas consists of 1 of charcoal and 1 of hydrogen united. No oxygen can be present in olefiant gas; because during the electrification it would be detected, either in the form of water or carbonic oxide.

It will be proper now to see how far the weights of the gases entering into combination, agree with the previous determinations. An atom of charcoal weighs 5.4 (see page 382), and 1 of hydrogen weighs 1, together making an atom of olefiant gas, 6.4. This atom will require 3 of oxygen for its combustion ; namely, 2 for the charcoal, to form carbonic acid, and 1 for the hydrogen, to form water ;

these 3 weigh 21; whence 6.4 parts of olefiant gas by weight, should take 21 of oxygen. Now supposing, according to Dr. Henry's result, that 100 measures of olefiant gas require 284 for their combustion; and further, that the specific gravity of oxygenous gas is 1.10 (agreeably to Allen and Pepys, as also Biot and Arago), we shall have 284 × 1.1 = 312.4, the weight of the oxygen; hence, if 21 : 6.4 : : 312.4 : 95, the weight of 100 measures of olefiant gas, corresponding to a specific gravity of .95. Hence, then, it appears that the weight of the gases combined, perfectly corroborates the above conclusions respecting the constitution of olefiant gas.

There are some remarkable circumstances attending the combustion of olefiant gas in Volta's eudiometer, which deserve notice as part of the history of the gas, but particularly as they put the constitution of the gas beyond all doubt. If 100 measures of oxygen be put to 100 of olefiant gas, and electrified, an explosion ensues, not very violent; but instead of a diminution, as usual, there is a great increase of gas; instead of 200 measures, there will be found about 360; some traces of carbonic acid are commonly observed, which disappear on passing two or three times through lime water; there will then remain, perhaps,

350 measures of permanent gas, which is all combustible, yielding by an additional dose of oxygen, carbonic acid and water, the same as if entirely burnt in the first instance. What, therefore, is this new gas in the intermediate state? The answer is clear. It is carbonic oxide and hydrogen mixed together, an equal number of atoms of each. One third of the oxygen requisite for the complete combustion, suffices to convert the carbone into carbonic oxide, and the hydrogen at the instant is liberated; hence the other two thirds are employed, the one to convert the carbonic oxide into acid, the other to convert the hydrogen into water. In fact, the 350 measures consist of nearly 170 of each gas, which together require nearly 170 of oxygen for their combustion.*

* M. Berthollet contends, that all the combustible gases into which carbone and hydrogen enter contain also oxygen: he calls them *oxycarburetted hydrogen.* Mr. Murray also enters into his views in this respect.—As far as relates to olefiant gas, it will be time enough for animadversion on this opinion, when the accuracy of the above facts and observations are questioned. But there is one circumstance which M. Berthollet has not explained in regard to this gas, and it turns upon a point which he and I acknowledge, but which is not perhaps generally received; namely, that *when two gases unite to form a third, this last is specifically heavier than the lighter of the two.* Now, in the above

The diameter of an atom of olefiant gas is .81 to hydrogen 1. And 100 measures of it contain as many atoms as 188 of hydrogen, or as 94 of oxygen, or (probably) as 200 of oxymuriatic acid ; whence the union of this last with olefiant gas, must be 2 atoms of the gas with 1 of the acid.

2. *Carburetted Hydrogen.*

The gas which I denominate carburetted hydrogen, was known in a state of mixture, to Dr. Priestley ; he called all such mixtures by the name of *heavy inflammable air.* Lavoisier, Higgins, Austin, Cruickshanks, Berthollet, Henry and others, have since cultivated this department of science.—Cruickshanks contributed much to unveil the subject, by pointing out carbonic oxide as an inflammable gas, *sui generis*, but often found mixed with other gases. No correct notion of the constitution of the gas about to be described, seems to have been formed till the atomic

instance, we find olefiant gas and oxygenous gas, uniting to form a third (according to his opinion), which is lighter by one half nearly than the lighter of the two. How is this new oxycarburetted hydrogen to be reconciled with the above principle ?

theory was introduced and applied in the investigation. It was in the summer of 1804, that I collected at various times, and in various places, the inflammable gas obtained from ponds; this gas I found always contained some traces of carbonic acid and a portion of azote; but that when cleared of these, it was of a uniform constitution. After due examination, I was convinced that just one half of the oxygen expended in its combustion, in Volta's eudiometer, was applied to the hydrogen, and the other half to the charcoal. This leading fact afforded a clue to its constitution.

Carburetted hydrogen gas may be obtained in a pure state, with the above exceptions, from certain ponds in warm weather. Clayey ponds, in the vicinity of a town, where soot and other carbonaceous matter is deposited, abound with this gas. The bottom of the pond being stirred with a stick, large bubbles ascend, which may be caught by filling a tumbler with water, and inverting it over the ascending bubbles. This gas is obtained nearly pure also by distilling pitcoal with a moderate red heat. It is now largely used as a substitute for lamps and candles, under the name of *coal gas*. According to Dr. Henry's analysis, coal gas does not usually contain more than 4 or 5 per cent of carbonic acid, sulphuretted hydrogen, and olefiant gas. The rest is principally carburetted hydrogen, but mixed with some atoms of carbonic oxide and hydrogen. The last portion of gas driven off from pit coal, seems to be entirely carbonic oxide and hydrogen. The distillation of wood and of moist charcoal, and many other vegetable substances, produces carburetted hydrogen, but highly charged with carbonic acid, carbonic oxide and hydrogen; the two last gases always appear exclusively at the end of the process.

The properties of carburetted hydrogen are, 1. It is unfit for respiration, and for the support of combustion. 2. Its specific gravity when pure, from my experience is very near 6. Dr. Henry finds the coal gas to vary from .6 to 78; but then the heaviest contain 15 per cent. of the heavy gases, carbonic acid, sulphuretted hydrogen, and olefiant gas.—Water absorbs $\frac{1}{27}$th of its bulk of this gas.—If 100 measures of carburetted hydrogen be mixed with 100 measures of oxygen (the least that can be used with effect), and a spark passed through the mixture, there is an explosion, without any material change of volume: after passing a few times through lime water, it is reduced a little, manifesting signs of carbonic acid. This residue is found to possess all the characters of a mixture of equal volumes of carbonic oxide and hydrogen. Upon adding 100 measures of oxygen to this residue and passing a spark, nearly 100 measures of carbonic acid are produced, and the rest of the produce is water. If 100 measures of carburetted hydrogen be put to upwards of 200 of oxygen, and fired over mercury, the result will be a diminution of near 200 measures, and the residuary 100 measures will be found to be carbonic acid.

Though carburetted hydrogen is naturally produced in many coal mines, and occasionally mixing with common air, exhibits some dreadful explosions in the large way; yet when mixed with common air, in Volta's eudiometer, it does not explode by a spark, unless the gas be to the air, as 1 to 10 nearly, and then feebly.

When a portion of carburetted hydrogen gas is electrified for some time, it increases in volume, in the end almost exactly doubling itself; at the same time a quantity of charcoal is deposited. The whole of the gas is then found to be pure hydrogen.

All these facts being compared, there cannot remain the least doubt as to the constitution of carburetted hydrogen. It is a compound of one atom of charcoal and two of hydrogen; the compound atom occupies the same space (nearly) as an atom of hydrogen ; and 4 atoms of oxygen are necessary to its complete com bustion; namely, 2 for the charcoal to form carbonic acid, and 2 for the hydrogen to form water. This conclusion derives a very elegant confirmation, from the facts observed by exploding the gas with one half of the oxygen requisite for complete combustion. In this case, each atom of the gas requires only 2 atoms of oxygen; the one joins to one of hydrogen and forms water; the other joins to the carbone to form carbonic oxide, at the same moment the remaining atom of hydrogen springs off. Thus there becomes 100 measures of carbonic oxide and 100 of hydrogen, or the same bulk as the original mixture.

As the weight of an atom of charcoal is 5.4, and 2 atoms of hydrogen are 2, the compound atom weighs 7.4; but as there are the same number of atoms of hydrogen and of carburetted hydrogen in the same volume, 7.4 represents the number of times that carburetted hydrogen is heavier than hydrogen. Now, the weight of common air is about 12 times as great as hydrogen; therefore, the relative weights or specific gravities of the two gases, are as 7 4 to 12, or as 6 to 1, nearly, which agrees with experience; hence we derive this conclusion, that carburetted hydrogen consists *entirely* of hydrogen and carbone, the whole weight of the gas being accounted for in the carbonic acid and water formed by its combustion.*

I think it proper to observe, that, according to my most careful experiments, 100 measures of this gas require rather more than 200 mea-

* According to M. Berthollet (Mem. d'Arcueil, tome 2d) the gas from charcoal is a triple compound of carbone, oxygen, and hydrogen. Whatever our speculative chemists may believe, no practical chemist in Britain adopts this idea. That it always contains more or less of oxygen no one disputes; but then the oxygen is united solely to the carbone forming carbonic oxide. The rest of the mixture consists of carburetted hydrogen and hydrogen. I never find any difficulty in ascertaining the relative quantities of each of the gases in such mixtures. For instance, suppose we take the first of Berthollet's nine specimens.

100 gas,	sp. gr. .462	took 81 oxy.	gave 56 carb. acid.	
20 carb. hyd.	sp. gr. .6	takes 42 ——	gives 21 ——	
34 carb. ox.	—— .94	—— 16	—— 32 ——	
46 hyd.	—— .08	—— 23	—— — ——	
100 mixt	—— .476	takes 81 ——	gives 53 ——	

Here it appears, that 20 measures of carb. hyd. + 34 carb. oxide + 46 hydrogen, constitute a mixture of 100 measures of the sp. grav. .476, which being burned, take 81 oxygen, and give 53 carb. acid. Hence this mixture may be considered as agreeing with Berthollet's gas from charcoal above specified.

sures of oxygen, and give rather more than 100 carbonic acid; but the difference is not more than 5 per cent. and may in general be neglected.—Hence, then, we may conclude that the diameter of an atom of carburetted hydrogen is nearly equal to that of hydrogen, but rather less.

SECTION 8.

HYDROGEN WITH SULPHUR.

There are two compounds of hydrogen with sulphur; the one, a well known elastic fluid denominated *sulphuretted hydrogen*, the other a viscid, oily compound, called *supersulphuretted hydrogen.* The former of these consists of 1 atom of each element,* the the latter probably of 1 atom of hydrogen united to 2 of sulphur.

1. *Sulphuretted Hydrogen.*

The best way I have found to obtain sulphuretted hydrogen in a pure state, is to heat a piece of iron to a white or welding heat in a

* The figure for sulphuretted hydrogen, plate 4, part 1, is incorrect: it ought to be 1 atom of hydrogen instead of 3, united to 1 of sulphur.

smith's forge, then suddenly drawing it from the fire, apply a roll of sulphur; the two being rubbed together, unite and run down in a liquid form, which soon fixes and becomes brittle. This compound or sulphuret of iron, is to be granulated and put into a gas bottle, to which dilute sulphuric acid is to be added, after which the gas comes over plentifully. When the sulphuret of iron is made in a crucible from iron filings and sulphur, it seldom answers well; it often gives hydrogen mixed with the sulphuretted hydrogen. The reason seems to be, that several sulphurets of iron exist; namely, the first, the second, the third, &c. and it is the second only, or that which is constituted of 1 atom of iron and 2 of sulphur, formed in the process above described, which is essential to the formation of sulphuretted hydrogen. The others either give hydrogen or no gas at all.

Sulphuretted hydrogen is unfit for respiration and for supporting combustion: it has a disagreeable smell, resembling that of rotten eggs; its specific gravity is 1.10 according to Kirwan, and 1.23 according to Thenard. Mr. Davy, I understand, makes it about 1.13. Some trials of mine a few years since, gave a result near Thenard's; but till a more correct one can be obtained, we may adopt the mean 1.16. Wa ter absorbs just its bulk of this gas; when, therefore, it is mixed with hydrogen, this last will be left after washing in water, or what is still better, in lime water. Sulphuretted hydrogen burns with a blue flame. When mixed with oxygen, in the ratio of 100 measures to 50 of oxygen (which is the least effective quantity), it explodes by an electric spark; water is produced, sulphur is deposited, and the gases disappear. If 150 or more measures of oxygen are used, then after the explosion over mercury, about 87 measures of sulphurous acid are found in the tube, and 150 of oxygen disappear, or enter into combination with both the elements of the gas.

From the experiments of Austin, Henry, &c. it has been established, that sulphuretted hydrogen undergoes no change of volume by electrification, but deposits sulphur. I have repeated these experiments, and have not been able to ascertain whether there was increase or diminution. The residue of gas is pure hydrogen.

From these facts, the constitution of sulphuretted hydrogen is clearly pointed out. It is 1 atom of sulphur and 1 of hydrogen, united in the same volume as 1 of pure hydrogen. When burned, 2 atoms of oxygen unite to 1 of sulphur to form sulphurous acid, and 1 of oxygen to 1 of hydrogen to form water. The weights of the elements confirm this constitution. One atom of sulphur has been found to weigh 13 (see page 393), to which adding 1 for hydrogen, we obtain the weight of an atom of sulphuretted hydrogen = 14; this number likewise expresses the number of times that sulphuretted hydrogen should exceed hydrogen in specific gravity. But common air exceeds hydrogen 12 times; therefore, 12 : 14 : : specific gravity of common air : sp. gravity of sulphuretted hydrogen = 1.16, agreeably to the preceding determination. Hence this gas is wholly composed of sulphur and hydrogen, as above.

Sulphuretted hydrogen unites, like the acids, to alkalies, earths, and metallic oxides, forming with them salts of definite proportions, which are called *hydrosulphurets*. Some of these are important chemical agents; but they are apt to undergo changes by keeping, especially in solution.

2. *Supersulphuretted Hydrogen.*

This compound may be obtained as follows. Let half an ounce of flower of sulphur and as much hydrate of lime, be gently boiled together in a quart of rain water for one hour more water may be added as it evaporates. After cooling, a clear yellow liquid is obtained, which is a solution of sulphuret of lime: it will vary in specific gravity from 1.01 to 1.02, according to circumstances.—To 6 ounces of this liquid put half an ounce of muriatic acid, and stir the mixture. In a short time, the mixture exhibits a milky appearance, and this becomes interspersed with brown oily dots, which gradually subside into an adhesive mass of a semiliquid form at the bottom. The liquid may then be poured off, and the brown mass washed with water, which is to be poured off. From 20 to 40 grains of this brown oily substance will be obtained; it is supersulphuretted hydrogen.

Scheele, Berthollet, and Proust, have made observations on this compound. When exposed to the air, or even in water, it exhales sulphuretted hydrogen, especially if warm. On account of its viscidi y and adhesiveness, it is very difficult to subject it to experience. If a portion of it touch the skin, &c. it requires a knife to scrape it off. It may be poured from one vessel to another by means of water, which prevents its adhering to the vessel. When a little of it is appplied to the tongue, a sensation of great heat, and a bitter taste are felt; the saliva becomes white like milk.

When liquid alkali is poured upon supersulphuretted hydrogen, heat is produced, hydrosulphuret is formed, and sulphur precipitated. —These facts have all been observed by me; though few if any of them are new.

There is no doubt this substance is formed of sulphur and hydrogen. I took 30 grains, and exposed them to a moderate heat in a glass capsule, over a candle, till they ceased to exhale sulphuretted hydrogen. The residuum weighed 21 grains; it was soft like clay; when ignited, it burned away with a blue flame, and left no sensible residuum. When it is considered, that supersulphuretted hydrogen is from the moment of its formation exhaling sulphuretted hydrogen, we cannot wonder that a portion of it should give less than half its weight of this gas. But scarcely any doubt can be raised that the sulphur of the gas is originally equal to that left behind; or that supersulphuretted hydrogen is constituted of 2 atoms of sulphur and 1 of hydrogen, and consequently weighs 27 times as much as hydrogen.

Though it is not our present business to explain the previous process by which the article under discussion is obtained; yet, as it will be some time before it comes regularly in our way, it may perhaps be allowable. Hydrate of lime, is 1 atom of lime and 1 of water united ; when boiled with sulphur as above, it takes 3 atoms of sulphur. The compound is *sulphuret of hydrate of lime.* When muriatic acid is mixed with it, the acid seizes the lime. The 3 atoms of sulphur divide the atom of water in such sort, that two of them take the hydrogen to form *supersulphuretted hydrogen,* and one takes the oxygen to form *sulphurous oxide.* This last occasions the milkiness of the liquid ; by long digestion the milkiness vanishes ; the sulphurous oxide is changed into sulphuric acid and sulphur, which last falls down, and forms nearly one fourth of that which originally existed in the sulphuret.

SECTION 9.

HYDROGEN WITH PHOSPHORUS.

There is only one combination of hydrogen with phosphorus yet known ; it is a gas denominated *phosphuretted hydrogen.* This gas may be procured as follows : Let an ounce or two of hydrate of lime (dry slacked lime) be put into a gas bottle or retort, and then a few small pieces of phosphorus, amounting to 40 or 50 grains. If the materials are sufficient to fill the bottle, no precaution need be used; but if not, the bottle or retort should be previously filled with azotic gas, or some gas not containing oxygen, in order to prevent an explosion. The heat of a lamp is then to be applied, and a gas comes which may be received over water. This gas is phosphuretted hydrogen; but sometimes mixed with hydrogen. —Liquid caustic potash may be used instead of hydrate of lime, in order to prevent the generation of hydrogen.

Phosphuretted hydrogen gas has the following properties: 1. When bubbles of it come into the atmosphere, they instantly take fire; an explosion is produced, and a ring of white smoke ascends, which is phosphoric acid: 2. It is unfit for respiration, and for supporting combustion: 3. Its specific gravity is .85, common air being denoted by unity: 4. Water absorbs $\frac{1}{27}$th of its bulk of this gas: 5. If the gas be electrified, the phosphorus is thrown down, and there finally remains the bulk of the gas of pure hydrogen. In fact, the phosphorus is easily thrown down, either by electricity, by heat, or by being exposed to a large surface of water. In this respect, phosphuretted hydrogen is nearly related to sulphuretted hydrogen.

Though phosphuretted hydrogen explodes

when sent into the atmosphere in bubbles, yet if sent into a tube of three tenths of an inch diameter, it may be mixed with pure oxygen, without any explosion. In all the experiments I have made, which are more than 20, I never had an instance of a spontaneous explosion. In this case, an electric spark produces a most vivid light, with an explosion not very violent; phosphoric or phosphorous acid and water are produced.

My experiments on the combustion of this gas give the following results: When 100 measures of pure phosphuretted hydrogen are mixed with 150 of oxygen, and exploded, the whole of both gases disappears; water and phosphoric acid are formed; when 100 measures of the gas are mixed with 100 oxygen, and fired, the whole of both gases still disappears; in this case, water and phosphorous acid are formed; when 100 measures are mixed with less than 100 of oxygen, phosphorous acid and water are formed, but part of the combustible gas remains unburnt.

As this gas is liable to be contaminated with hydrogen, sometimes largely, on account of the facility it possesses of depositing phosphorus, it is expedient to ascertain the exact proportion of phosphuretted hydrogen to hydrogen in any proposed mixture. This I find may easily be done. Whenever a sufficient quantity of oxygen is afforded, the whole of the combustible gas is consumed : The exact volume of oxygen and its purity must be noted ; the quantity of oxygen in the residue must also be noted. Then the total diminution after the explosion, being diminished by the oxygen consumed, leaves the combustible gas. Now, as phosphuretted hydrogen takes $1\frac{1}{2}$ times its bulk of oxygen, and hydrogen takes $\frac{1}{2}$ its bulk of oxygen ; we shall obtain the following equations, if P denote the volume of phosphuretted hydrogen, H that of hydrogen, O that of oxygen, and $S = P + H$, the whole of the combustible gas.

$$P = O - \tfrac{1}{2}S$$
$$H = 1\tfrac{1}{2}S - O$$

From these equations, the ratio of the two gases in any mixture is deduced. The analysis may be corroborated as follows : To any mixture containing a certain volume of phosphuretted hydrogen, let the same volume of oxygen be added ; after the explosion, the diminution will be just twice the volume of oxygen. In this case, the phosphuretted hydrogen is preferred by the oxygen ; phosphorous acid and water are formed, and the hydrogen remains in the tube. If more oxy gen is put than the phosphuretted hydrogen; then the diminution after firing is more than twice the oxygen.

The investigation respecting the proportion of hydrogen mixed with phosphuretted hydrogen, was instituted chiefly in consequence of a difference of opinion respecting the specific gravity of the latter gas. I had found 100 cubic inches to weigh about 26 grains; Mr. Davy informed me he had found 100 inches to weigh only 10 grains: the difference is enormous. I requested Dr. Henry would assist me in repeating the experiment. We obtained a gas, such that 100 inches weighed 14 grains; this result surprized me; but upon burning the gas with oxygen, it was found only to take its bulk of that gas, and consequently to be half hydrogen and half phosphuretted hydrogen, which satisfactorily explained the difficulty. Mr. Davy's gas, I conceive, must have been $\frac{1}{3}$ phosphuretted hydrogen and $\frac{2}{3}$ hydrogen, at the time it was weighed; however this may be, it is evident, from what is related above, that nothing certain can be inferred relative to the specific gravity of this gas, unless a portion of the gas be analyzed previously to its being weighed; a circumstance of which I was not at first sufficiently aware.

I have recently procured some phosphuretted hydrogen gas from caustic potash and phosphorus; an accident prevented me obtaining a sufficient quantity to weigh; but I got 5 or 6 cubic inches, which of course were mixed with the azotic gas previously put into the retort. The pure combustible gas was of such character, that 100 measures required only 85 of oxygen for their combustion; it was consequently 35 phosphuretted hydrogen and 65 hydrogen per cent., and probably would have weighed after the rate of 10 or 11 grains for 100 cubic inches. I expected much purer gas.

As to the constitution of phosphuretted hydrogen, it is clearly 1 atom of phosphorus united to 1 of hydrogen, occupying the same space as 1 of elastic hydrogen. In combustion, the atom of hydrogen requires one of oxygen, and the atom of phosphorus requires one or two of oxygen, according as we intend to produce phosphorous or phosphoric acid. Hence it is that 100 measures of phosphuretted hydrogen require 50 oxygen to burn the hydrogen, 50 more of oxygen to form phosphorous acid, and 50 more to form phosphoric acid. The weight of the gas corroborates this conclusion: it has been seen that the atom of phosphorus weighs nearly 9 (page 415); this would make the specific gravity of phosphuretted hydrogen equal to 10 times that of hydrogen, which it actually is, or nearly so, from the foregoing experiments.

The next compounds to be considered in course, would be those of *azote* with *carbone*, with *sulphur*, and with *phosphorus*; but such compounds either cannot be formed, or they are yet unknown.

SECTION 10.

CARBONE WITH SULPHUR, WITH PHOSPHORUS, AND SULPHUR WITH PHOSPHORUS.

1. *Carbone with Sulphur.*

In the 42d vol. of the An. de Chimie, page 136, Clement and Desormes have announced a combination of carbone and sulphur, which they call *carburetted sulphur*. They obtain it by sending the vapour of sulphur over red hot charcoal; it is collected in water in the form of an oily liquid of the specific gravity 1.3. This liquid is volatile, like ether, expanding any gas into which it is admitted, and forming a permanent elastic fluid over the mercury of a barometer. No gas is produced at the same time as the liquid. When too much sulphur is driven through, instead of a liquid, a solid compound is formed which crystallizes in the tube. They seem to have shewn that the compound does not contain sulphuretted hydrogen.—In the 64th vol. of the Journal de Physique, A. B. Berthollet endeavours to prove that the liquid in question is a compound of hydrogen and sulphur, and contains no charcoal. The facts adduced are not sufficient to decide the question either way. I should be unwilling to admit, with Clement and Desormes, that the two inelastic elements, charcoal and sulphur, would form an elastic or volatile compound ; yet, they have rendered it highly probable that charcoal makes a part of the compound, as it disappears during the process. I think it most probable, that Berthollet is correct in the idea that this liquid contains hydrogen. We know of no other volatile liquid that does not contain hydrogen. Perhaps it will be found a triple compound of hydrogen, sulphur, and charcoal.

2. *Carbone with Phosphorus.*

A combination of carbone and phosphorus has been pointed out by Proust, in the 49th volume of the Journal de Physique, which he names *phosphuret of carbone.* It is the reddish substance which remains when new made phosphorus is strained through leather in warm water. The proportion of the two elements has not been ascertained.

3. *Sulphur with Phosphorus.*

Melted phosphorus dissolves and combines with sulphur, and that in various proportions, which have not yet been accurately ascertained. The compounds may be denominated *sulphurets of phosphorus.* The method of forming these compounds, is to melt a given weight of phosphorus in a tube nearly filled with water, and then to add small pieces of sulphur, keeping the tube in hot water, taking care not to exceed 160°, or 170°, or 180°, because the new compound begins to decompose water rapidly at those high temperatures. Pelletier has given us some facts towards a theory of these various combinations, in the

4th vol. of the An. de Chimie. He found that a mixture of sulphur and phosphorus remained fluid at a much lower temperature than either of them individually; and that different proportions gave different fusing or congealing points. One part of phosphorus, combined with $\frac{1}{8}$th of sulphur, congealed at 77°; one part with $\frac{1}{4}$, at 59°; one part with $\frac{1}{2}$, at 50°; one part with 1, at 41°; one part with 2, at 54°$\frac{1}{2}$; but a certain portion was fluid, and the rest solid; and one part with 3, at 99°.5.

One would be apt to think, from these experiments, that sulphur and phosphorus might be combined in all proportions; but the observation on the 5th led me to suspect that it might have been applied to some others if the results had been carefully noted.—I mixed 18$\frac{1}{2}$ grains of phosphorus and 13 of sulphur in a graduated tube, put in water, and immersed the whole into water of 160°. The phosphorus having been rendered fluid as usual, at 100°, gradually reduced the sulphur, till the whole assumed a liquid form of the specific gravity 1.44. It remained uniformly fluid at 45°, but was wholly congealed at 42°. Here were two atoms of phosphorus united to one of sulphur. I then added 6$\frac{1}{2}$ grains of sulphur, making the mixture 18$\frac{1}{2}$ phosphorus, and 19$\frac{1}{2}$

sulphur; this new mixture was reduced to uniform fluidity at 170°, and was of 1.47 specific gravity; reduced to 47°, one part was fluid and the other solid, the latter being at the bottom of the tube. This solid part was not completely reduced to fluidity in the temperature 100°. This seems to indicate that two distinct combinations took place; the one, two atoms of phosphorus and one of sulphur, liquid at 47°; the other, one atom of phosphorus and one of sulphur, solid under 100°. I next added 6½ grains more of sulphur, making in the whole 18½ phosphorus and 26 sulphur, consequently in such proportion as to afford a union of one atom of each; the union was completed in a temperature of 180°: the specific gravity was 1.50. Cooled down to 80°, the whole was solid; heated to 100°, the whole became a semi-liquid, uniform mass. Being afterwards heated to 140°, the whole became fluid; but upon cooling again, the greatest part congealed at 100°, but ⅓d or ¼th remained liquid down to 47°.—From these experiments, it is most probable that one atom of each forms a combination which is solid at 100° or below; but that being heated, it is apt to run into the other mode of combination, or that constituted of two atoms of phosphorus and one of sulphur. The properties of these two species of sulphuret of phosphorus I have not had an opportunity to investigate. The water in the tube is evidently decomposed in part by the compound ; it becomes milky, probably through the oxide of sulphur, and both sulphuretted and phosphuretted hydrogen seem to be formed in small quantities at temperatures above 160°.

SECTION 11.

FIXED ALKALIES.

The fate of the two fixed alkalies, potash and soda, has been rather remarkable. They had long been suspected to be compound elements, but no satisfactory proof was given. At length Mr. Davy, by his great skill and address in the application of galvanism to produce chemical changes, seemed to have established the compound nature of these elements, both by analysis and synthesis. They appeared to be *metallic oxides*, or peculiar metals united to oxygen. Consistent with this idea, some account of the metals, denominated *potasium* and *sodium*, has been given in this work. (See page 260). But from what follows, it appears most probable, that these metals are compounds of potash and soda with hydrogen, and that the two fixed alkalies still remain among the undecompounded bodies.

1. *Potash.*

Potash is obtained from the ashes of burned wood. Water dissolves the saline matter of the ashes, and may then be poured off and evaporated by artificial heat: the salt called *potash* remains in the vessel. If the salt so obtained be exposed to a red heat, it loses combustible matter, becomes white, and is in part purified: in commerce it is then called *pearl-ash.* This mass is still a mixture of various salts, but is constituted chiefly of *carbonate of potash.* In order to obtain the potash separate, let a quantity of *pearl-ash* (or what is still better, *salt of tartar* of the shops, which is this pearl-ash reduced almost to pure carbonate of potash) be mixed with its weight of water, and the mixture be stirred; after the undissolved salt has subsided, pour off the clear solution into an iron pan, and mix with it a portion of hydrate of lime, half the weight of the liquid; then add a quantity of water equal to the weight of the ingredients, and boil the mixture for several hours, occasionally adding more water to supply the waste. When the liquid is found not to effervesce with acids, the ebullition may be discontinued. After the lime has subsided, the clear liquid is to be decanted, and then boiled down in a clean iron pan till it assumes a viscid form, and acquires almost a red heat. It may then be poured into molds, &c. and it immediately congeals. The substance so obtained is potash nearly pure; but it still contains a considerable portion of water, some foreign salts, oxide of iron, and frequently some unexpelled carbonic acid. The water may amount to 20 or 25 per cent. upon the whole weight, and the other substances to 5 or 10 per cent. In this process, the carbonic acid of the potash is transferred to the lime.

If potash of still greater purity be required, the method practised by Berthollet may be pursued. The solid potash obtained as above must be dissolved in alcohol; the foreign salts will fall to the bottom insoluble; the liquid solution may then be decanted into a silver bason, the alcohol be evaporated, and the fluid potash exposed to a red heat. It may be poured out upon a clean polished surface, where it instantly congeals into solid plates of potash, which are to be broken and put into well stoppered bottles, to prevent the access of air and moisture. This potash is a solid, brittle, white mass, consisting of about 84 parts potash and 16 water, in 100 parts, and is the purest that has ever yet been obtained.

Potash may be exhibited in a more regular crystalline form by admitting more water to it. If the solution be reduced to the specific gravity of 1.6, or 1.5, upon cooling, crystals will be formed, containing about 53 per cent. of water, or more, if the air is cold. These crystals are called *hydrate of potash.* Hence solid hydrate of potash may be formed, containing from 84 per cent. of potash to 47, or under.

Potash has a very acrid taste; it is exceedingly corrosive if applied to the skin, so as to obtain the name of *caustic.* The specific gravity of the common sticks of potash used by surgeons, I find to be 2.1; but these are a mixture of potash and carbonate of potash, with 20 or 30 per cent. of water. If pot-ash were obtained pure, I apprehend its specific gravity would be about 2 4.

When crystals of potash (that is, the hydrate) are exposed to heat, they become liquid, the water is gradually dissipated with a hissing noise, till at length the fluid acquires a red heat. It then remains tranquil for some time; but if the heat be increased, white fumes begin to arise copiously. The alkali and water both evaporate in this case; therefore, the process cannot be used to expel the last portion of water from the alkali. If the hydrate be taken in the red hot and tranquil state, it contains 84 per cent. potash and 16 water. This is ascertained by saturating a given weight of it with sulphuric acid, when sulphate of potash is formed free from water, and 100 parts of the hydrate give only 84 parts to the new compound.

Water has a strong affinity for potash. If a portion of the 84 per cent. hydrate be put into as much water, great heat is immediately produced, equal to that of boiling water. But it is observable that the crystallized hydrate containing much water, when mixed with snow, produces excessive cold. When potash is exposed to the air, it attracts moisture and carbonic acid, becoming a liquid carbonate. Potash dissolved in water, and kept in a stoppered bottle, retains its causticity: it is called *alkaline ley*, and may be had of various strengths and specific gravities.

Potash, and the other alkalies, change vegetable colours, particularly blues, into green. —Potash is of great utility in the arts and manufactures, particularly in bleaching, dying, printing, soap and glass manufactures. It unites with most acids to form salts. It does not unite with any of the simple substances, as far as is yet known, except hydrogen, and that in a circuitous way, as will presently be noticed. The hydrate of potash unites with sulphur; but the compound, consisting of three or more principles, cannot yet be discussed.

The theory of the nature and origin of potash still remains in great obscurity. The great question, whether it is a constituent principle of vegetables, or formed during their combustion, is not yet satisfactorily answered. One circumstance is favourable to the investigation of the nature of potash, the weight of its ultimate particle is easily ascertained; it forms very definite compounds with most of the acids, from which it appears to be 42 times the weight of hydrogen. The following proportions of the most common salts with base of potash, are deduced from my experience: they are such that good authorities may be found both for greater and less proportions of the different elements.

	per cent.		
Carbonate of potash,	31.1, acid	+ 68.9 base,	as 19 : 42
Sulphate ———	44.7 ——	+ 55.3 ——	34 : 42
Nitrate ———	47.5 ——	+ 52.5 ——	38 : 42
Muriate ———	34.4 ——	+ 65 6 ——	22 : 42

The above salts are capable of sustaining a red heat, and may therefore be supposed to be free from water; at all events, the potash must contain the same quantity of water in combination with the respective acids, as appears from the uniformity of its weight. The above numbers, 19, 34, 38 and 22 represent, as the reader will recollect, the weights of the atoms of the respective acids, except the nitric, which is double. As water has so strong an affinity for potash, and as the weight of the elementary particle of potash above deduced is more than five times that of water, it may still be supposed that water enters into the constitution of potash, or that it is compounded of some of the lighter earths with azote, oxygen, &c. From present appearances, however, the notion that potash is a simple substance seems more probable than ever.

From the above observations, it appears that potash ought still to be considered as a simple substance, and would require to be placed among such substances, but that it cannot be obtained alone. In that state which approaches nearest to purity it is a hydrate, containing at least 1 atom of water united to 1 of potash, amounting to 16 per cent. of water. This hydrate is therefore a *ternary* compound, or *one* of three elements, and ought to be postponed till the next chapter: but, in the present state of chemical science, utility must be allowed in some instances to supersede methodical arrangements. The fixed alkalies are most useful chemical agents, and the sooner we become acquainted with them the better; more especially, as some of the first chemists of the present age have been led into considerable mistakes, by presuming too much upon their knowledge of the nature and properties of these familiar articles.

In the *Memoires de l'Institut de France*, 1806, Berthollet published researches on the laws of affinity, from which some extracts are given in the Journal de Physique for March 1807.—By these, it appears that he found sulphate of barytes to consist of 26 acid and 74 base, and sulphate of potash of 33 acid and 67 base. The former of these results was corroborated by the previous experience of Thenard; but both are so remote from the uniform results of other chemists, that they could never be generally adopted. At length Berthollet discovered the error, and has announced it in the 2d vol. of the Memoires d'Arcueil. It consisted in mistaking the hydrates of barytes and potash for pure barytes and potash. It seems to have been generally adopted, but certainly prematurely, that barytes and potash, in a state of fusion, were pure, or free from water. But upon due investigation, he found that fused potash contains 14 per cent. of water: my experience as well as theory, leads me to adopt 16 per cent of water, which accords with the position of 1 atom of each of the elements uniting to form the hydrate; namely, 42 by weight of potash with 8 of water. This discovery reconciles the jarring results on the proportions of the above neutral salts, and throws light upon some other interesting subjects of chemical analysis.

2. *Hydrate of Potash.*

Upon turning my attention to this subject, I soon perceived the want of a table exhibiting the relative quantities of potash and water in all the combinations of these two elements. In a state of solution, the specific gravity may be taken as a guide; but this is not quite so convenient when the compound is in a solid form. I found nothing of the kind in any publication, and therefore undertook a course of experiments to determine the relative quantities of potash, &c. in the various solutions. The results are contained in the following table, which I would have to be considered only as an approximation to truth; but it will certainly have its use till a more complete and accurate one be obtained. Dr. Henry was so obliging as to facilitate my progress, by presenting me with portions of the fixed alkalies, prepared after Berthollet's method.

Table of the quantity of real potash in watery solutions of different specific gravities, &c.

Atoms. Potash Water	Potash per cent. by weight.	Potash per cent. by measure.	Specific gravity.	Congealing point.	Boiling point.
1+ 0	100	240	2.4	unknown.	unknown
1+ 1	84	185	2.2	1000°	red heat
1+ 2	72.4	145	2.0	500°	600°
1+ 3	63.6	119	1.88	340°	420°
1+ 4	56.8	101	1.78	220°	360°
1+ 5	51.2	86	1.68	150°	320°
1+ 6	46.7	75	1.60	100°	290°
1+ 7	42.9	65	1.52	70°	276°
1+ 8	39.6	58	1.47	50°	265°
1+ 9	36.8	53	1.44	40°	255°
1+10	34.4	49	1.42		246°
	32.4	45	1.39		240°
	29.4	40	1.36		234°
	26.3	35	1.33		229°
	23.4	30	1.28		224°
	19.5	25	1.23		220°
	16.2	20	1.19		218°
	13.	15	1.15		215°
	9.5	10	1.11		214°
	4.7	5	1.06		213°

Remarks on the Table.

The first column contains the number of atoms of potash and water in the several combinations to 10 atoms of water: the weight of an atom of potash is taken to be 42, and 1 of water 8. From these data the second column is calculated. There did not appear any striking characteristic of distinction between the first, second, third, &c. hydrates, (if they may be so called) except that the first bears a red heat in the liquid form, with tranquillity and without loss of weight. Before this, the water is gradually dissipated with a hissing noise and fumes. I remarked, however, that when a solution of potash is boiled down till the thermometer indicates upwards of 300°, the evaporation of the water, and the rise of the thermometer, are desultory; that is, the operations appear somewhat stationary for a time, and then advance quickly; how far this may arise from the nature of the compound, or from the imperfect conducting power of the liquid in those high temperatures, I could not determine without more frequent repetitions of the experiment.

The third column is, as usual, obtained by multiplying the second column by the specific gravity; it is often more convenient in practice to estimate quantity by measure than by weight.

The fourth column denotes the specific gravity; below 1.60 the hydrate is completely fluid, or may be made so by a moderate heat, but above that temperature, I found some difficulty in ascertaining the specific gravity, and was obliged sometimes to infer it from the tenor of the table. The common sticks of potash of the druggists are of the sp. gr. 2.1, which I found by plunging them into a graduated tube filled with mercury, and marking the quantity that overflowed. These sticks are a mixture of hydrate and carbonate. Real potash must, I conceive, be heavier than they are. The relation of the second and fourth columns was ascertained by taking a given weight of the alkaline solution, saturating it with test sulphuric acid (1.134), and allowing 21 grains of alkali for every 100 measures of acid (containing 17 real) which the alkali required.

The 5th column denotes the temperatures at which the different hydrates congeal or crystallize. This part of the subject deserves much more accurate enquiry than I have been able to bestow upon it. No doubt the different hydrates might be distinguished this way. Proust talks of a crystallized hydrate of potash, containing 30 per cent. of water; and Lowitz of one containing 43 per cent. of water. They calculate, I presume, upon the supposition of fused potash being free from water; if so,

Proust's hydrate is the fourth of our table, and Lowitz's the sixth. I would not have much trust to be put in the temperatures I have marked in this column.

The sixth column indicates the temperatures at which the different specific gravities boil. This is easily ascertained, except for the high degrees, in which an analysis of the hydrate was required upon every experiment. I believe the results will be found tolerably accurate. As the range of temperature is large, this may be found a very convenient method of ascertaining the strength of alkaline solutions, when the specific gravities are unknown.

3. *Carbonate of Potash.*

Though it be premature to enter into the nature of carbonate of potash, a triple compound, yet its utility as a test is such as to require it to be noticed in the present section. Indeed it may generally be a substitute for the hydrate of potash, and it can much more readily be procured in a state of comparative purity. The carbonate I mean is that which consists of one atom of acid united to one of potash, which by some writers is called a *subcarbonate.* It is, of course, constituted of 19 parts of acid by weight united to 42 of potash. This salt is to be had in tolerable purity of the druggists, under the name of *salt of tartar*: but when it is to be used in solution for pure carbonate, a large quantity of the salt, and a small quantity of water, are to be mixed and agitated; then let the undissolved salt subside, and pour off the clear solution, which may be diluted with water, &c.

This salt is well known to be, like the dry hydrate of potash, very deliquescent. I took 43 grains of carbonate of potash that had just before been made red hot, put them into a glass capsule exposed to the air; in one day the weight became 50 grains; in three days, 61 grains; in seven days, 75 grains; in 11 days, 89 grains; in 21 days, 89+ grains; in 25 days, 90 grains. The specific gravity was 1.54 nearly. All the water is, however, driven off by a moderate heat; namely, that of 280°. It supports a high red heat before fusion, and when fused loses no weight, remaining without sublimation, and undecompounded. I ascertained that it was a perfect carbonate, by dissolving 61 grains of pure dry salt in lime water, when 42 grains of carbonate of lime were thrown down, corresponding to 19 grains of carbonic acid.

Table of the quantity of real carbonate of potash in watery solutions of different specific gravities.

Atoms. Carb. of Pot. Water	Carb. Potash per cent. by weight.	Carb. Potash per cent. by measure.	Specific gravity.	Boiling point.
1+ 0	100	260	2.60	280°
1+ 1	88.4	212	2.40	265°
1+ 2	79.2	170	2.15	258°
1+ 3	71.8	140	1.95	252°
1+ 4	65.6	118	1.80	247°
1+ 5	60.4	103	1.70	244°
1+ 6	56	91	1.63	241°
1+ 7	52.1	82	1.58	238°
1+ 8	48.8	75	1.54	235°
1+ 9	45.8	69	1.50	232°
1+10	43.3	63	1.46	229°
	41.7	60	1.44	227°
	39	55	1.41	225°
	36.2	50	1.38	222°
	33.6	45	1.34	220°
	30.5	40	1.31	218°
	27.3	35	1.28	217°
	24	30	1.25	216°
	20.5	25	1.22	215°
	16.8	20	1.19	214°
	13.2	15	1.15	214°
	9	10	1.11	213°
	4.7	5	1.06	213°

This table is similar in structure to the preceding. The first column contains the number of atoms of water joined to one of carbonate of potash, which last weighs 61. The second contains the weight of carbonate of potash per cent. in the compound, and the third the grains of carbonate in 100 water grain measures of the compound, found by multiplying the numbers in the second and fourth columns together. The fourth contains the specific gravities; the relations of these to the quantities in the second column were found, by taking a given weight of the solution, and saturating it with a certain number of measures of test sulphuric acid (1.134), allowing 21 real potash, or $30\frac{1}{2}$ carbonate, for every 100 measures of acid required; because such acid contains 17 per cent. by measure of real sulphuric acid, and that requires 21 of potash.

The strongest solution of this salt that can be obtained is of the specific gravity 1.54. This consists of 1 atom of carbonate and 8 of water; but by putting dry carbonate into that solution, various mixtures may be formed up to the specific gravity 1.80; above that the specific gravity is scarcely to be obtained but by inference. I could not obtain a solid stick of fused carbonate but what was spongy, I suppose from incipient decomposition. It may be observed, that the specific gravity 1.25, which contains 30 per cent of carbonate, is that which I prefer as a test for acids; because the solution contains 21 per cent. pure potash, and 100 measures of it consequently require 100 measures of the test acids.

I found a specimen of the pearl-ash of commerce to contain 54 parts carbonate of potash,

22 parts of other salts, and 24 parts of water in the hundred.

The fifth column denotes the temperature at which the saline solutions boil. This will be found generally a good approximation to truth. I observed the thermometer did not rise above 280° as long as any visible moisture remained; as soon as that vanished, the salt assumed the character of a hard and perfectly dry substance.

In the course of these experiments, I took a quantity of carbonate of potash, and heated it red hot; then weighed it; after which I put to it as much water as afforded 1 atom to 1; namely, 8 parts water to 61 salt. The salt was then pulverized in a mortar; it was put out upon white paper, and appeared a white, dry salt; but upon pouring it back into the mortar, some particles of the salt adhered to the paper. The same quantity of water was again put to it. Upon mixing them with a knife, the whole mass assumed a pasty consistence, and adhered to the knife in the shape of a ball; after being well rubbed in the mortar, it again assumed a white, dry appearance. Upon paper, it seemed like salt of tartar some time exposed to the air. Several particles stuck to the paper, but were easily removed by a knife. The addition of another atom of water reduced the compound to the consistence of bird-lime; but after standing it cut like half dried clay. The next atom of water reduced it to the consistence of bookbinders paste. The fifth atom of water reduced it to a thick fluid, consisting of dissolved and undissolved salt. This, by the successive application of like portions of water, became a perfect fluid with 8 atoms of water to 1 of carbonate of potash. Its specific gravity was 1.5; but there was some undissolved sulphate of potash subsided, the salt of tartar not having been previously purified.

4. *Potasium, or Hydruret of Potash.*

Since writing the articles on Potasium and Sodium (page 260 and seq.), and the subsequent articles on fluoric and muriatic acid (page 277 and seq.), a good deal more light has been thrown on these subjects. Two papers on the subjects have been published by Mr. Davy; a series of essays by Gay Lussac and Thenard, are contained in the 2d vol. of the Memoires d'Arcueil; the same volume also contains a paper by Berthollet, announcing an important discovery relating to the fixed alkalies; namely, that in a state of fusion by heat, they contain a definite proportion of water in chemical combination. Upon reconsidering the former facts, and comparing them with the more recent ones, I am obliged to adopt new views respecting the nature of these new metals. Mr. Davy still adheres to his original views, and which indeed were the only rational ones that could be formed (supposing the fused alkalies to contain no water), namely, that potash is the oxide of potasium; Gay Lussac and Thenard, on the contrary, consider potash as undecompounded, and potasium a compound of hydrogen and potash, analogous to the other known compounds of hydrogen and elementary principles. This last is the only one, I think, that can be admitted either from synthetic or analytic experiments, so as to be reconcileable with the facts; but I do not coincide with all the conclusions which the French chemists have deduced. Mr. Davy has furnished us with the most definite and precise facts; and though I was led to controvert some of them (see page 289 and seq.), it was principally through my having adopted his views of the nature of potasium: I am now persuaded those results were more accurate than I imagined.

Mr. Davy first attempted to decompose the fixed alkalies, by applying Voltaic electricity to saturated watery solutions ; in this case, oxygen and hydrogen gas were obtained, evidently proceeding, as he concluded, from the decomposition of the water. But when any potash that had previously been fused, was substituted for the watery solution, no hydrogen gas was given out at the negative pole, but potasium was formed, and pure oxygen was given out at the positive pole. The residual potash was unaltered. The conclusion he drew was, that the potash was decomposed into potasium and oxygen. But it now appears, that fused potash is composed of 1 atom of water and 1 of potash. The electricity operates upon this last atom of water to separate its elements ; it succeeds in detaching the atom of oxygen, but that of hydrogen draws the atom of potash along with it, forming an atom of potasium. The atom of hydrate weighing 50 (= 42 potash + 8 water) is decomposed into one of potasium, weighing 43, and one of oxygen weighing 7. Hence the atom of potasium is composed of 1 potash + 1 hydrogen, weighing 43 ; and not of 1 potash — 1 oxygen, weighing 35, as stated at page 262.

The method of obtaining potasium, discovered by the French chemists, is to find the first hydrate of potash in a state of vapour over red hot iron turnings, in an iron tube intensely heated; hydrogen gas is given out, potasium is formed and condensed in a cool part of the tube, and part of the potash is found united to the iron. In this mode of producing potasium, its constitution is not so obvious as in the former. The two methods, however, together, shew that fused potash contains both oxygen and hydrogen, which is now abundantly confirmed by experiments of a different kind. It seems probable that in the latter method the hydrate of potash is partly decomposed into potash and water, and partly into potasium and oxygen; in both cases the iron acquires the oxygen.

The specific gravity of potasium is .6, or .796, according to Davy; but .874 according to Gay Lussac and Thenard. The levity of it, combined with its volatility at a low red heat, agrees with the notion of its being potash and hydrogen, or *potassetted hydrogen*, resembling the other known compounds of sulphur, phosphorus, charcoal, arsenic, &c. combined with hydrogen.

When burned in oxygen gas, potasium produces potash as dry as possible to be procured, according to Mr. Davy; that is, the first hydrate. When potasium is thrown into water it burns rapidly, decomposing the water, and giving off hydrogen. Calculating the oxygen from the quantity of hydrogen, Mr. Davy finds 100 (hydrate of) potash contain from 13 to 17 oxygen: Gay Lussac seems to make it 14. For, 2.284 grammes of potasium gave 649 cubic centimetres of hydrogen; reduced, 35.5 grains gave 34.5 cubic inches English measure, which correspond to 17.25 inches of oxygen = 5.9 grains. Hence 35.3 + 5.9 = 41.2 grains of hydrate; and 41.2 : 5.9 : : 100 : 14. But this is exactly the quantity that theory would assign; for, 43 potasium + 7 oxygen = 50 hydrate, which gives just 14 oxygen in the hundred.

Potasium burns spontaneously in oxymuriatic acid gas; muriate of potash is formed, and probably water. It decomposes sulphuretted, phosphuretted, and arseniuretted hydrogen gas, according to Gay Lussac and Thenard, and unites to the sulphur, &c. with some of the hydrogen. Mr. Davy finds tellurium to unite with the hydrate of potash by Voltaic electricity without decomposing it. Potasium burns in nitrous gas and nitrous oxide, forming dry hydrate of potash, and evolving azote. It burns in sulphurous and carbonic acid, and in carbonic oxide; hydrate of potash which unites to the sulphur is formed, or hydrate of potash and charcoal.

The combustion of potasium in muriatic acid gas is particularly worthy of notice. Both Mr. Davy and the French chemists agree that when potasium is burned in muriatic acid gas, muriate of potash is formed, and hydrogen evolved, which agrees in quantity with that evolved in the decomposition of water by the same quantity of metal. But, what is most astonishing, they both adopt the same explanation, when their different views of the constitution of potasium require them to be opposite. Mr. Davy had two ways in which he might account for the phenomenon; the one was to suppose that a part of the acid was decomposed, and furnished the oxygen to the metal to form the oxide (potash), which joined to the remainder of the acid, and the hydrogen was an evolved elementary principle of that part of the acid decomposed; and the other, to suppose that the acid gas contained in a state of union just as much water as was sufficient to oxidate the metal (this would have been thought an extraordinary circumstance a few years ago) Either of these positions was consistent; but he adopted the latter, and seemed to confirm it by shewing that a given quantity of muriatic acid gas afforded the same quantity of muriate of silver, whether combined previously with potash or potasium.

This explanation did not meet my views as well as the former. I endeavoured to account for the facts (page 289) on the notion of a decomposition of the acid. Two circumstances conspired to incline me to this view: The one was, that hydrogen seemed on other accounts to be a constituent of muriatic acid; the other was, that water does not appear in any other instance to be combined with any elastic fluid; I mean in such way that if the water be removed, the rest of the molecule will carry along with it the character of the whole. In one respect I mistook the data, having overrated the weight of muriatic acid gas.—I would now be understood to abandon the explanation founded on the decomposition of the acid; and to adopt the much more simple one that the muriatic acid combines with the potash of the potasium, at the same instant expelling the hydrogen; in this way there is no occasion for any water either combined or otherwise. It exceeds my comprehension how Gay Lussac and Thenard should insist so largely on the opinion that muriatic acid gas contains water, and that principally, as it should seem, in order to account for the hydrogen evolved during the combustion of potasium, and the supposed oxydation of the metal.

It has been stated that potasium burns in silicated fluoric acid gas (page 283), the result is fluate of potash and some hydrogen. The theory of this is not obvious.

Potasium acts upon ammoniacal gas. Mr. Davy found that when 8 grains of the metal were fused in ammoniacal gas, between $12\frac{1}{2}$ and 16 cubic inches of the gas were absorbed, and hydrogen evolved corresponding to the oxydation of the metal by water, that is, 1 atom of hydrogen for 1 atom of potasium. The new compound becomes of a dark olive colour. By applying a greater degree of heat the ammonia is in part expelled again; but part is also decomposed. Gay Lussac and Thenard say, that by admitting a few drops of water to the compound, the whole of the elements of the ammonia are recoverable, and nothing but caustic potash remains. Mr. Davy affirms the results of the decomposition to be somewhat different. It seems pretty evident, that in this process two atoms of ammonia unite to one of potasium, expelling its hydrogen at the same moment. For, 43 grains of potasium would require 12 of ammonia; and therefore 8 will require $2\frac{1}{4}$ grains, which correspond to $12\frac{1}{2}$ cubic inches.

5. *Soda.*

Soda is commonly obtained from the ashes of plants growing on the sea-shore, particularly from a genus called *salsola*; in Spain, where this article is largely prepared, it is called *barilla*. In Britain, the various species of *fucus* or sea-weed are burnt, and their ashes form a mixture containing some carbonate of soda; this mixture is called *kelp*. Soda is found in some parts of the earth combined with carbonic acid, and in others combined with muriatic acid, as minerals; and hence it has been called the *fossil* or *mineral* alkali, to distinguish it from potash or the *vegetable* alkali.

To obtain soda in as pure a state as possible, recourse must be had to a process similar to that for obtaining potash. Pure carbonate of soda must be treated with hydrate of lime and water; the carbonate of soda is decomposed; the soda remains in solution in the liquid, the carbonic acid unites to the lime, and the new compound is precipitated. Afterwards the clear liquid must be decanted and boiled down; the water gradually goes off with a hissing noise till the soda acquires a low red heat, when the alkali and remaining water become a tranquil liquid. This liquid may be run out into molds, &c. when it instantly congeals into a hard mass, and is then to be preserved in bottles for use. If still greater heat be applied, the alkali and water are together dissipated in white fumes.

Soda thus obtained is a solid, brittle, white mass, consisting of about 78 parts pure soda and 22 water per cent.; according to d'Arcet (Annales de Chimie, Tome 68, p. 182) the alkali is only 72; but I believe that is too low. With more water, soda may be had in crystals, like potash, probably containing 50 or 60 per cent. of water. Soda, like potash, is extremely caustic; it is deliquescent, and produces heat when dissolved in water. The specific gravity of fused soda I find to be 2, by pouring it into a graduated glass tube. There is some reason to apprehend that pure soda, could it be obtained, would be specifically heavier than potash, though its ultimate particle is certainly of less weight than that of the latter. The properties and uses of soda are much the same as those of potash; indeed, the two alkalies were long confounded, on account of their resemblances. The compounds into which they enter are in many instances essentially different, and the weights of their atoms are very unequal. The origin of soda in vegetables is somewhat obscure, though it may be derived from the muriate of soda in the water of the sea.

The weight of an atom of soda is easily derived from the many definite compounds which it forms with the acids; it appears to be 28 times that of hydrogen. The carbonate, sulphate, nitrate and muriate of soda, are all well known salts. From a comparison of my own experiments with those of others on the proportions of these salts, free from water, I deduce the following:

	per cent.			
Carbonate of soda	40.4 acid,	+ 59.6 base,	as	19 : 28
Sulphate ——	54.8 ——	45.2 ——		34 : 28
Nitrate ——	57.6 ——	42.4 ——		38 : 28
Muriate ——	44 ——	56. ——		22 : 28

These proportions scarcely differ 1 per cent. from those of Kirwan and other good authorities. The numbers 19, 34, 38 and 22 being the weights of the respective atoms of acids, the number 28 must be the weight of an atom of soda. Hence we find that soda is a peculiar element, differing from every one we have yet determined in weight. From the weight of the element soda, it may be suspected to be a compound of water, oxygen, or some of the lighter elements; but from present appearances, no such suspicion seems well founded. Soda should then, with propriety, be treated as an elementary principle. We shall proceed to the hydrate, the carbonate, and the hydruret of soda, for reasons which have been given under the head of potash.

6. *Hydrate of Soda.*

Soda, in what has till lately been considered its pure state, is combined with water. The smallest portion of water seems to be one atom to one of soda ; that is, 8 parts of water by weight to 28 of soda, or 22 per cent. of water. I have not obtained soda purer than that of d'Arcet of 72 per cent. ; but it always contained some carbonic acid and other impurities, which incline me to conclude that 78 per cent. would be the highest attainable purity ; this may be called the first hydrate : it is hard and brittle, and twice the weight of water. The second, third, fourth, and fifth hydrates are, I apprehend, crystalline ; but my experience does not warrant me to decide upon their nature ; the sixth, and those with more water, are all liquid at the ordinary temperature ; their specific gravity is obtained in the usual way, and the corresponding quantity of real alkali is ascertained by the test acids.

The following Table for soda, is constructed after the manner of that for potash (page 476).

It will be found moderately accurate; but I could not give it the attention it deserves. Nothing of the kind has been published to my knowledge; yet, such tables appear to me so necessary to the practice of chemical enquiries, that I have wondered how the science could be so long cultivated without them.

That solution which will be found most convenient for a test, is of the specific gravity 1.16 or 1.17, and contains 14 per cent. by measure of real alkali; consequently, 100 measures require the same volume of acid tests for their saturation.

Table of the quantity of real soda in watery solutions of different specific gravities, &c.

Atoms. Soda. Water.	Soda per cent. by weight.	Soda per cent. by measure.	Specific gravity.	Congealing point.	Boiling point.
1 + 0	100	230?	2.30?	unknown.	unknown
1 + 1	77.8	156	2.00	1000°	red hot
1 + 2	63.6	118	1.85	500°	600°
1 + 3	53.8	93	1.72	250°	400°
1 + 4	46.6	76	1.63	150°	300°
1 + 5	41.2	64	1.56	80°	280°
1 + 6	36.8	55	1.50		265°
	34	50	1.47		255°
	31	45	1.44		248°
	29	40	1.40		242°
	26	35	1.36		235°
	23	30	1.32		228°
	19	25	1.29		224°
	16	20	1.23		220°
	13	15	1.18		217°
	9	10	1.12		214°
	4.7	5	1.06		213°

7. *Carbonate of Soda.*

The salt I call *carbonate of soda*, is to be had of the druggists in great purity, under the name of purified sub-carbonate of soda. It is obtained in the form of large crystals, containing much water; but when exposed to the air for some time, these crystals lose most of their water, and become like flour. I took 100 grains of fresh crystallized carbonate of soda, and exposed it to the action of the air in a saucer: In 1 day it was reduced to 80 grains; in 2 days, to 64 grains; in 4 days, to 49 grains; in 6 days, to 45 grains; in 8 days, to 44 grains; and in 9 days it was still 44 grains, had the appearance of fine dry flour, and probably would have lost no more weight. It was then exposed to a red heat, after which it weighed 37 grains nearly. Now, it is a well established fact, that the common carbonate of soda, heated red, is constituted of 19 parts of acid and 28 of soda; or 40.4 acid and 59.6 base, per cent. nearly. Klaproth says, 42 acid, 58 base; Kirwan says, 40.1 acid, 59.9 base. It is equally well established that the crystallized carbonate recently formed in a low temperature, contains about 63 per cent. water, as above determined. All experience confirms this; Bergman and Kirwan find 64 parts of water, Klaproth 62, and d'Arcet 63.6. Hence the constitution of the crystallized carbonate is easily ascertained; for, if 37 : 63 : : 47 (= 19 + 28) : 80, the weight of water attached to each atom of the carbonate; that is, 10 atoms of water unite to 1 of carbonate of soda to form the common crystals. Again, if 47 : 8 : : 37 : 6.3 = the weight of water attached to 37 parts of carbonate of soda, to correspond with 1 atom of water; but 37 + 6.3 = 43.3; from this it appears that 100 parts of crystallized carbonate being reduced to 44 or 43.3, indicates that all the 10 atoms of water are evaporated, except one. It should seem, then, that the ordinary efflorescence of this salt is not dry carbonate, but 1 atom of carbonate and 1 of water. This supposition is confirmed by experience; for, in 5 days the above 37 grains of heated carbonate became 44 grains by exposure to the air.

There is another very remarkable character of the carbonate of soda, which, however, I apprehend will be found to arise from a general law in chemistry; when a quantity of common crystallized carbonate is exposed to heat in a glass retort, as soon as it attains a temperature about 150°, it becomes fluid as water; but when this fluid is heated to 212°, and kept boiling a while, a hard, small grained, salt is precipitated from the liquid, which, upon examination, I find to be the *fifth* hydrate, or one atom of carbonate of soda united to 5 atoms of water. For, 100 grains of this salt lose 46 by a red heat; but 1 atom of carbonate weighs 47, and 5 atoms of water weigh 40, together making 87; now, if 87 of such salt contain 40 water, 100 will contain 46.—The clear liquid resting upon the fifth hydrate has the specific gravity 1.35; on cooling, the whole liquid crystallizes into a fragile, icy mass, which dissolves with a very moderate heat. This appears by the test acid to be constituted of 1 atom of carbonate and 15 atoms of water. Thus the tenth hydrate, by heat, is resolved and converted into the fifth and fifteenth; in like manner, probably, the fifteenth might be transformed into the tenth and thirtieth hydrate. When any solution below 1.35 sp. gravity is set aside to csystallize, the fifteenth hydrate is formed in the liquid, and finally the residuary liquid is reduced to the sp. gravity of 1.18. By treating this liquid solution with the test acids, it will be found to consist of 1 atom of carbonate to 30 of water. It is of course that solution which the common crystals of carbonate always form, when duly agitated with water; or a *saturated* solution at the mean ordinary temperature of the atmosphere. By heat, other liquid solutions may be obtained from 1.85 to 1.35; but they soon crystallize; such may be called *supersaturated* solutions.

The different species of hydrates in crystals have different specific gravities, as might be expected; that of the fifteenth is 1.35; that of the tenth is 1.42, and that of the fifth 1.64. These were found by dropping the crystals into solutions of carbonate of potash till they were suspended, or by weighing them in saturated solutions of the same. I could not ascertain that of the pure carbonate and the first hydrate.

When carbonate of soda is used for a test alkali, the specific gravity 1.22 would be that solution which contains 14 per cent. by measure of alkali, of which 100 measures would require 100 of test acid for saturation; but, as that solution cannot be preserved without partial crystallization, it will be better to substitute a solution of half the strength; namely, that of 1.11; then 200 measures of the solution will require 100 of test acid.

The following Table contains the characters of various combinations of carbonate of soda and water, resulting from my investigations.

Table of the quantity of real carbonate of soda in watery compounds of different specific gravities.

Atoms. Carb. Soda. Water.	Carb. Soda per cent. by weight.	Carb. Soda per cent. by measure.	Specific gravity.	Congealing point.	Boiling point.
1 + 0	100	200?	2.00?	unknown.	unknown.
1 + 1	85.5	162?	1.90?	——	——
1 + 5	54	89	1.64	——	——
1 +10	37	52.5	1.42	150°	
1 +15	28.8	39	1.35	80°	220°
1 +20	22.7	28	1 26		217°
1 +30	16.4	19.5	1.18		214°
		15	1.15		——
		10	1.10		213°
		5	1.05		——

The state of the carbonates in the above table it may be proper to notice. The pure carbonate is in the state of a dry powder ; so is the first hydrate, not to be distinguished in appearance from the pure carbonate. The fifth hydrate may be obtained in a crystalline mass, by heating the common carbonate till a proper portion of water is driven off. Its specific gravity is then easily found. The tenth hydrate is the common carbonate of the shops in crystals. The fifteenth hydrate may be had either in a liquid or solid form, as has been observed. The twentieth hydrate is a liquid without any remarkable distinction that I have discovered. It is liable to partial crystallization. The thirtieth hydrate is a liquid, being the saturated solution at common temperature; this would probably wholly crystallize at no very reduced temperature. The 2d, 3d, 4th, 6th, &c. hydrates, I have not found to offer any remarkable discrimination.

8. *Sodium, or Hydruret of Soda.*

According to the present state of our knowledge, the account of sodium given at page 262, will require some modification. As the article from which sodium has always been obtained is the first hydrate of soda, and as in the electrization of fused hydrate of soda, no gas is given out, according to Mr. Davy, but oxygen; it follows of course that sodium must be a compound of soda and hydrogen, which may be called a hydruret of soda. Mr. Davy, conceiving soda in a state of fusion to be pure or free from water, as was the common opinion at the time, concluded that in the electrization of it the soda was decomposed into sodium and oxygen. This conclusion does not now appear to be tenable, though Mr. Davy still adheres to it, without having shewn what becomes of the water acknowledged to be present in every instance of the formation of sodium and potasium (Philos. Trans. 1809), to the amount of 16 per cent. upon the compound.

Though Mr. Davy's original method of obtaining sodium by Voltaic electricity is the most instructive, as to the nature of the new product, yet, that of Gay Lussac and Thenard is the most convenient when a quantity of the article is required. That is, to pass the vapour of red hot hydrate of soda over iron turnings in a gun barrel, heated to whiteness. The hydrate seems to be decomposed in two ways; in part it is resolved into sodium, or hydruret of soda, and oxygen, the former of which distils into a cooler receptacle of the barrel, and the latter unites to the iron; in part, the hydrate is decomposed into water and soda, and the former again into oxygen, which unites to the iron, and hydrogen which escapes, whilst the soda unites to the iron or its oxide, forming a white metallic compound.

The specific gravity of sodium is stated by Mr. Davy at .9348. The weight of its ultimate particle (being 1 atom of soda and 1 of hydrogen) must be 29, and not 21, as stated at page 263. Consequently, 100 parts of the first hydrate of soda, or fused soda, contain 80.6 sodium and 19.4 oxygen per cent. This agrees with that one of Mr. Davy's experiments which gave the least portion of oxygen.

Sodium amalgamates with potasium, according to Gay Lussac and Thenard, in various proportions, and the alloys are more fusible than either of the simple metals, being in some cases liquid at the freezing point of water. In general, the properties of sodium are found to agree with those of potasium so nearly, as not to require distinct specification.

SECTION 12.

EARTHS.

The class of bodies called *earths* by chemists are nine in number; their names are *Lime*, *Magnesia*, *Barytes*, *Strontites*, *Alumine* or *Argil*, *Silex*, *Yttria*, *Glucine* and *Zircone*. The three last are recently discovered and scarce.

The earths constitute the bases of the fossil kingdom. Though they have frequently been suspected to be compound bodies, and several attempts have been made to decompose them, it does not yet appear but that they are simple or elementary substances. Some of the earths possess alkaline properties; others are without such properties; but they all partake of the following characters: 1. They are incombustible, or do not unite with oxygen ; 2. they are inferior to the metals in lustre and opacity ; 3. they are sparingly soluble in water ; 4. they are difficultly fusible, or resist great heat without alteration ; 5. they combine with acids ; 6. they combine with each other, and with metallic oxides ; and, 7. their specific gravities are from 1 to 5.

The latest attempt to decompose the earths is that of Mr. Davy ; he seems to have shewn, that some of the earths are analogous to the fixed alkalies, in respect to their properties of forming metals ; but these metals, like those of the alkalies, are most probably compounds of hydrogen and the respective earths.

1. *Lime.*

This earth is one of the most abundant ; it is found in all parts of the world, but in a state of combination, generally with some acid. When united with carbonic acid, it exists in large strata or beds in the form of chalk, limestone, or marble ; and it is from some of these that lime is usually obtained.

The common method of obtaining lime, is to expose pieces of chalk or limestone in a kiln for a few days to a strong red or white heat ; by this process, the carbonic acid is driven off, and the lime remains in compact masses of nearly the same size and shape as the limestone, but with the loss of $\frac{9}{20}$ths of its weight. It is probable, the intermixture of the limestone and coal in the combustion of the latter contributes, along with the heat, to the decomposition. The lime from chalk is nearly pure; but that from common limestone contains from 10 to 20 per cent. of foreign substances, particularly alumine, silex, and oxide of iron.

Lime thus obtained, which is commonly called *quicklime*, is white and moderately hard, but brittle. Its specific gravity, according to Kirwan, is 2.3. It is corrosive to animal and vegetable substances; and, like the alkalies, converts coloured vegetable infusions, particularly blue, into green. It is infusible. It has a strong attraction for water, so as to rob the atmosphere of its vapour; when exposed to the atmosphere, it gradually imbibes water, and in a few days falls down into a fine white dry powder; in this process, if pure, it acquires 33 per cent. in weight; after this, it begins to exchange its water for carbonic acid, and carbonate of lime is slowly regenerated. When 1 part of water is thrown upon 2 of quicklime, the lime quickly falls to powder with intense heat, calculated to be 800° (page

89) ; this operation is called *slaking* the lime, and is preparatory to most of its applications ; the new compound is denominated *hydrate of lime*, and appears to be the only proper combination that subsists between lime and water. By a red heat the water is driven off and the lime remains pure.

As lime combines with the principal acids hitherto considered, and forms with them perfectly neutral salts ; and as the proportions of these salts have been experimentally ascertained with precision, we are enabled to determine the weights of an atom of lime : thus,

	Acid.		Base.		
Carbonate of lime,	44	+	56	per cent. as	19 : 24
Sulphate ——	58.6	+	41.4	——	34 : 24
Nitrate ——	61.3	+	38.7	——	38 : 24
Muriate ——	47.8	+	52.2	——	22 : 24

Carbonate of lime is, I believe, universally allowed to contain either 44 or 45 per cent. of acid ; and sulphate is mostly supposed to contain 58 per cent. acid, the extremes being 56 and 60. The proportions of the other two salts have not been so carefully determined , but it is easy to satisfy one's self that the proportions assigned are not wide of the truth. Let 43 grains of chalk be put into 200 grain measures of the test nitric acid (1.143), or the test muriatic (1.077), and it will be found that the lime will be wholly dissolved, and the acids saturated. Hence it follows that the elementary atom of lime weighs 24. I have formerly stated it at 23, supposing carbonate of lime to be, according to Kirwan, 45 acid + 55 lime per cent. The difference is scarcely worth consideration; but experience seems to warrant 24 rather than 23 for the atom of lime.

When a large quantity of water is thrown upon a piece of quicklime, it sometimes refuses to slake for a time; perhaps this is caused by the water preventing the rise of temperature. In this case the water does not dissolve the lime; hence it should seem that lime properly speaking is not soluble in water; but hydrate of lime is readily soluble, though in a small degree. The solution is called *lime-water*, and is a very useful chemical agent.

Lime-water may be formed by agitating a quantity of hydrate of lime in water; distilled or rain water should be preferred. One brisk agitation is nearly sufficient to saturate the water; but if complete saturation is required, the agitation should be repeated two or three times. After the lime has subsided the clear liquid must be decanted and bottled for use. Authors differ as to the quantity of lime dissolved by water: some say that water takes $\frac{1}{500}$ of its weight of lime; others, $\frac{1}{600}$. The fact is, that few have tried the experiment with due care. Dr. Thomson, in the 4th ed. of his chemistry says, from his experience, $\frac{1}{758}$. This is much nearer the truth than the other two. One author says, that water of 212° takes up double the quantity of lime that water of 60° does, but deposits the excess on cooling: no experimental proof is given. If he had said *half* instead of double, the assertion would have been nearly true. I have made some experiments on this subject, and the results are worth notice.

When water of 60° is duly agitated with hydrate of lime, it clears very slowly; but a quantity of the lime-water may soon be passed through a filter of blotting paper, when it becomes clear and fit for use. I found 7000 grains of this water require 75 grains of test sulphuric acid for its saturation. Consequently it contained 9 grains of lime. If a quantity of this saturated water, mixed with hydrate of lime, be warmed to 130° and then agitated, it soon becomes clear; 7000 grains of this water decanted, require only 60 grains of test sulphuric acid in order to produce saturation. The same saturated lime-water was boiled with hydrate of lime for two or three minutes, and set aside to cool without agitation; it very soon cleared, and 7000 grains being decanted, required only 46 grains of test acid to be neutralized, the test acid being as usual 1.134. Hence we deduce the following table.

1 part water of	takes up of lime	takes up of dry hydrate of lime
60°	$\frac{1}{778}$	$\frac{1}{584}$
130°	$\frac{1}{972}$	$\frac{1}{729}$
212°	$\frac{1}{1270}$	$\frac{1}{952}$

This table leads us to conclude that water at the freezing temperature would take nearly twice the quantity of lime that water at the boiling temperature takes; I had not an opportunity to try this in the season of these experiments; but I am informed the calico-printers find a sensible difference in lime-water in different seasons of the year, and that in winter it is most subservient to their purpose, and least so in summer. As water takes up so small a portion of lime, and cold water more than warm, one would suppose it was the effect of *suspension* rather than *solution*. With this view I tried whether the addition of a little gum to the water would not increase its solvent power; but the result was, that water of 60° took precisely the same quantity of lime, whether with or without gum. I found that a deep earthen vessel which had stood some months with lime-water exposed to the air, still contained $\frac{1}{800}$ of its weight of lime.

Lime-water has an acrid taste, notwithstanding the small quantity of lime. It operates on colours like the alkalies. Certain blue colours, such as syrup of violets, are changed to green; infusion of litmus, which has been converted from blue to red by a little acid, has its blue colour restored by lime-water, and archil solution, reddened by an acid, is restored to its purple colour by lime-water. When exposed to the air, lime-water has a thin crust formed on its surface; this is carbonate of lime, the acid being derived from the atmosphere; it is insoluble, and falls to the bottom; in time the whole of the lime is thus converted into carbonate, and the water remains pure. If a person breathes through a tube into lime-water, it is rendered milky through the formation of carbonate, or if water containing carbonic acid be poured into it; but a double quantity of the acid forms a supercarbonate of lime, which is soluble in a considerable degree. Though lime is soluble in water in so small a quantity, yet a portion of distilled water may be mixed with $\frac{1}{100}$ of its bulk of lime-water, and the presence of lime will be shewn by the test colours, or by nitrate of mercury, &c.

Lime combines with sulphur and with phosphorus: these compounds will be considered under the heads of sulphurets and phosphurets. It combines also with the acids, and forms with them neutral salts. Lime unites to certain metallic oxides, particularly those of mercury and lead; but the nature of these last compounds is not much known.

One of the great uses of lime is in the formation of mortar. In order to form mortar, the lime is slaked and mixed up with a quantity of sand, and the whole well wrought up into the consistence of paste with as little water as possible. This cement, properly interposed amongst the bricks or stones of buildings, gradually hardens and adheres to them so as to bind the whole together. This is partly, perhaps principally, owing to the regeneration of the carbonate of lime from the carbonic acid of the atmosphere. The best ingredients and their proportions to form mortar for different purposes, do not seem yet to be well understood.

2. *Magnesia.*

This earth is obtained from a salt now called *sulphate of magnesia*, which abounds in sea-water and in some natural springs. According to the best analyses, crystallized sulphate of magnesia consists of 56 parts of pure dry sulphate, and 44 parts water in the hundred. Some authors find more water in this salt; namely, from 48 to 53 per cent.; but Dr. Henry, in his analysis of British and foreign salt, in the Philos. Trans. 1810, takes notice of a crystallized sulphate of magnesia containing only 44 per cent. water; and the specimen of sulphate which I have had for many years bears the same character. I am, therefore, inclined to adopt this as the true proportion of water. Now, Dr. Henry found that 100 grains of the above sulphate of magnesia produced 111 or 112 grains of sulphate of barytes; and it is well established that $\frac{1}{3}$ of this last salt is acid; hence, the sulphuric acid in 100 sulphate of magnesia (56 real) is equal to 37 grains; consequently the magnesia is equal to 19 grains: but the weight of an atom of sulphuric acid is 34; therefore, 37 : 19 : : 34 : 17, nearly, which must be the weight of an atom of magnesia, on the supposition that sulphate of magnesia is constituted of one atom of acid united to one of base, of which there is no reason to doubt. I have in the first part of this work, page 219, stated the weight of magnesia to be 20; it was deduced chiefly from Kirwan's analysis of sulphate of magnesia; but from present experience I think it is too high. Though few of the salts of magnesia have been analyzed with great precision, yet the weight of the atom of magnesia derived from different analyses would not fall below 17, nor rise above 20. Dr. Henry and I analyzed the common carbonate of magnesia well dried in 100°, and found it to lose 40 per cent. by acids, and 57 per cent. by a moderate red heat. Hence it should consist of 43 magnesia, 40 carbonic acid, and 17 water. We found the carbonate begin to give out water and some acid about 450°; but it supported a heat of 550° for an hour without losing more than 16 per cent. Hence the carbonate must be constituted of 1 atom of acid, 1 of magnesia, and 1 of water, stating the magnesia at 20; for, 19 + 8 + 20 = 47; and if 47 : 19, 8, and 20 : : 100 : 40, 17 and 43 respectively, according to the above experiments. I have reason to think, however, that the weight of the atom of magnesia ought rather to be deduced from the sulphate than the carbonate; because it is probable that this last always contains a small portion of sulphate of lime, when prepared by the medium of common spring water; this portion will be found in the result of the analysis by fire, and will be placed to the account of magnesia.

Wherefore I conclude the weight of an atom of magnesia to be 17. It is said that a supercarbonate of magnesia is obtainable; but when sulphate of magnesia and supercarbonate of soda in solution are mixed together, there is a great effervescence and disengagement of carbonic acid, and nothing but the common carbonate of magnesia is precipitated according to my experience. Dr. Henry, indeed, obtained a crystallization by exposing a dilute mixture for some time; the crystals were small opake globules, about the size of small shot; but upon examination, they proved to be nothing but carbonate of magnesia united to 3 atoms of water instead of 1 atom. For, 100 grains lost 70 by a red heat, and 30 by acids; whence its constitution was 30 acid + 30 earth + 40 water, or 19 acid + 19 earth + 24 or 25 water. The constitution of crystallized sulphate of magnesia must, therefore, be 1 atom of acid + 1 atom of magnesia + 5 atoms of water; in weight 34 + 17 + 40 = 91; this gives per cent. 37 acid + 19 base + 44 water, agreeably to Dr. Henry's experience abovementioned.

The constitution of the most common salts of magnesia, in their dry state will, therefore, be as under;

	Acid.		Base.		
Carbonate of magnesia	53	+	47	per cent. as	19 : 17
Sulphate —— ——	66.7	+	33.3	—— ——	34 : 17
Nitrate —— ——	69	+	31	—— ——	38 : 17
Muriate —— ——	56.4	+	43.6	—— ——	22 : 17

The nitrate of magnesia in the above table agrees with that of Kirwan, and Richter, and the muriate with that of Wenzel.

To obtain magnesia, the sulphate must be dissolved in water, and a quantity of pure potash in solution must be added; the magnesia is then thrown down, and may be separated by filtration. Or if carbonate of potash be put into the solution of sulphate of magnesia, carbonate of magnesia will then be precipitated, which may be separated by filtration; this last must be exposed to a red heat to drive off the carbonic acid; the former need only to be dried in a gentle heat.

Magnesia is a white, soft powder, possessing little taste and no smell; its specific gravity is said to be 2.3. It operates on vegetable colours like lime and the alkalies. It is infusible by heat, and very sparingly soluble in water. According to Kirwan, it requires 7000 times its weight of water to dissolve it; I found it require 16,000 times its weight of water in one experiment. When exposed to the air, magnesia, like lime, attracts 1 atom of water to 1 of magnesia, amounting to about 47 per cent by my experience; it attracts carbonic acid but very slowly. It does not combine with any of the simple substances, except perhaps hydrogen and sulphur. With the acids it forms neutral salts, which are found frequently to combine with other salts.

As the sulphate of magnesia is the ordinary combination of this earth exhibited as a soluble salt, it may be of use to have a table shewing the quantity of real dry sulphate, and of ordinary crystallized sulphate, in given weights or measures of solutions of different specific gravities. The table is founded on my own experience.

Table of sulphate of magnesia.

Atoms. Mag. Water.	Dry sulphate of magnesia per cent. by weight.	Dry sulphate of magnesia per cent. by measure.	Common crystallized sulphate of mag. per cent. by measure.	Specific gravity.
1 + 0	100			
1 + 5	56	93	166	1.66 sol.
1 +18	44.4	66.6	119	1.50 liq.
1 +10	39	55.4	99	1.42
1 +15	30	39	69.6	1.30
		31	55	1.25
		24	42 8	1.20
		18	32.1	1.15
		12	21.4	1.10
		6	10.7	1.05

The fifth hydrate is the ordinary crystallized sulphate; the eighth is the strongest liquid solution obtained by boiling; and the fifteenth is a saturated solution at 60°.

3. *Barytes.*

The earth now denominated *barytes*, was discovered by Scheele in 1774. Since then the labour and experience of several distinguished chemists have added much to the knowledge both of the earth and its compounds; so that now it may perhaps be said to be the best understood of all the earths. It occurs most frequently in combination with sulphuric acid, the compound being called *sulphate of barytes*, formerly *ponderous spar*, and is found about mines, particularly of copper. It also occurs in combination with carbonic acid, though rarely; the compound is denominated *carbonate of barytes*.

Barytes may be obtained either from the sulphate or the carbonate. The former must be pulverized, mixed with charcoal, and exposed in a crucible to a red heat for some hours; the sulphate is thus changed into a sulphuret. This sulphuret is to be treated with nitric acid, when the sulphur is thrown down, and the barytes combines with the acid; the acid may then be driven off by a red heat, and barytes will remain in the crucible. If the carbonate be used, it must be pulverized, mixed with charcoal, and exposed for some time in a crucible to the heat of a smith's forge. Boiling water will then dissolve the pure barytes, leaving the charcoal and carbonate, and upon cooling, crystals of hydrate of barytes are obtained. The greatest part of the water may be driven off by heat.

Pure barytes obtained by the former method is a greyish white body, easily reduced to powder. It has a harsh and caustic taste, and if swallowed proves poisonous. Like lime, when exposed to the atmosphere, it absorbs water, and then parts with it for carbonic acid. It changes certain vegetable blues to green. Its specific gravity is nearly 4. Barytes forms various combinations with water, called *hydrates*, which will presently be mentioned. It combines with sulphur and phosphorus, but not with the other simple substance. The sulphuret and phosphuret will be considered under their respective heads. The weight of the ultimate particle of barytes can be very nearly approximated, and appears to be 68, or twice the weight of an atom of sulphuric acid. This appears from the following statement of the proportions of the most common barytic salts, which have been successfully investigated.

	Acid.		Base.		
Carbonate of barytes	22	+	78	per cent. as	19 : 68
Sulphate —— ——	33.3	+	66.7	—— ——	34 : 68
Nitrate —— ——	36	+	64	—— ——	38 : 68
Muriate —— ——	24.4	+	75.6	—— ——	22 : 68

The following respectable authorities agree in assigning 22 per cent. acid to carbonate of barytes; namely, Pelletier, Clement, Desormes, Klaproth, and Kirwan; and more recently Mr. Aikin finds 21.67, and Mr. James Thomson, 21.75 (Nicholson's Journal, vol. 22 and 23, 1809). The last mentioned chemist finds sulphate of barytes to be 33 acid, and 67 barytes. His conclusion corroborates the previous ones of Withering, Black, Klaproth, Kirwan, Bucholz, and Berthier, who all fix the acid at or near 33 per cent. Vauquelin, Rose, Berthollet and Thenard, and Clement and Desormes find 32 or more acid; and Fourcroy and Aikin, 34. It is very satisfactory to see the near coincidence in regard to the constitution of this salt; because it is frequently made a test of the quantity of sulphuric acid and of sulphur. Mr. J. Thomson finds 59.3 barytes per cent. in nitrate of barytes, Clement and Desormes 60, Kirwan 58 and 55 at different trials, and Fourcroy and Vauquelin 50. These results differ considerably from each other, and are all below the proportion assigned above; but it must be observed that crystallized nitrate of barytes contains water, and perhaps various quantities of water according to the temperature in which it crystallizes; now, if the atom of nitrate be associated with 1 atom of water, then the proportion of barytes per cent. will be 59.6, which nearly agrees with Thomson, and Clement and Desormes; if with 2 atoms of water, the barytes will be 55.7 per cent.; if with 3 atoms, then 52.3, &c.—Crystallized muriate of barytes appears clearly to consist of an atom of dry muriate + 2 atoms of water; or 22 acid + 68 barytes + 16 water; this reduced gives 20.8 acid + 64.1 barytes + 15.1 water per cent.—For, Kirwan finds 20 acid + 64 base + 16 water; Fourcroy, 24 acid + 60 base + 16 water; and Aikin, 22.9 acid + 62.5 base + 14.6 water per cent., which agree with each other, and with the theory as nearly as can be expected.

Barytes combines with most acids, and forms with them neutral salts. In many respects it appears to be related to the fixed alkalies, only in weight it is nearly the same as both of them put together.

Hydrate of Barytes.

When pure barytes, obtained from the nitrate by heat, is exposed to the air, or is moistened by water, it combines with it, and that in various degrees, forming a number of *hydrates*, which have not been sufficiently attended to and discriminated; much heat is evolved during the combination: it was mistaking the first hydrate of barytes for pure barytes that caused the uncertainty for some time in regard to the proportions of the elements of sulphate of barytes (see page 474). Now, if an atom of barytes weigh 68, the first hydrate will weigh 76, to which if 34 sulphuric acid be added, we shall have an atom of sulphate of barytes = 102, (for the water is driven off by the union of the acid and base); if then we conceived the hydrate to be pure barytes, we should conclude that 76 barytes united to 26 sulphuric acid to form 102 sulphate, which is very near the former mistaken conclusion of Thenard and Berthollet. Hence, then, there is reason to conclude that their barytes, kept some time in a red heat, was in reality the first hydrate, or one atom of barytes and one of water. When pure barytes is dissolved in boiling water, a solution is formed of specific gravity exceeding 1.2; on cooling, great part of it crystallizes; these crystals are the *twentieth* hydrate, or consist of 1 atom of barytes and 20 of water, or 30 barytes and 70 water per cent.; if they are exposed to a heat about 400° or 500°, they melt, great part of the water is dissipated, and a dry white powder is obtained, which is the *fifth* hydrate. In this operation, 228 parts ($= 68 + 20 \times 8$) are reduced to 108 ($= 68 + 5 \times 8$), or 100 to 47, which is exactly the reduction obtained experimentally by Dr. Hope. This dry powder melts below a red heat; but I have not been able to find what it would be reduced to by exposure to a red heat, because it acquires carbonic acid, even in a crucible, as Berthollet has observed, almost as fast as it loses water. My experience on the crystals of barytes has been limited; but from the following I conclude they are the *twentieth* hydrate. I took 80 grains of fresh crystallized barytes, and dissolved them in 1000 grains of water; the solution was of the specific gravity 1.024; this solution took 70 grain measures of test sulphuric acid to saturate it, and afforded 36 grains of dried sulphate of barytes: of this 12 grains were acid and 24 barytes. Whence we learn, 1st, that 80 grains of crystals are equal to 24 real barytes, or 228 equal to 68; but $228 = 20 \times 8 + 68$, which shews that 20 atoms of water are united to 1 of barytes; 2d, that the decimals in the second and third places of the expression for the specific gravity, denote the quantity of real barytes in 1000 grain measures of the solution. This last must evidently hold without any material error in all the inferior solutions; and hence the strength and value of barytic water may be known by its specific gravity, an advantage which does not practically appertain to lime-water. By subsequent trials, however, I found the quantity of barytes rather overrated.

The following sketch of a table of the hydrate of barytes may have its use, till a more ample and correct one can be constructed.

Table of the Hydrate of Barytes.

Atoms. Baryt. Water.	Barytes per cent. by weight.	Barytes per cent. by measure.	Specific gravity.		Congealing point.
1 + 0	100	400?	4.00?	sol.	unknown.
1 + 1	90	——	——	—	——
1 + 5	63	——	——	—	——
1 + 20	30	48	1.6	—	200°?
1 + 36	19	25	1.3	fl.	150°?
1 + 275	2.6	2.7	1.03	—*	40°?
	1.8	1.8	1.02	—	——
	0.9	0.9	1.01	—	——

4. *Strontites.*

The mineral from which this earth is obtained was first found in the lead-mine of Strontian in Argyleshire, Scotland. The earth and

* This is a saturated solution in the mean temperature of 60°.

its distinguishing properties, were pointed out by Dr. Hope in an essay read to the Royal Society of Edinburgh, in 1792, and published in their Transactions, 1794. Several distinguished chemists have since confirmed and extended these investigations. The Scotch mineral is a *carbonate* of strontites; but the earth has since been found in various parts combined with sulphuric acid.

Strontites is obtained from the sulphate or carbonate of strontites, by the same processes as barytes from the like compounds; indeed, it bears so close a resemblance to barytes, both in its free and combined state, as to have beeh confounded with it. Strontites has much the same acrid taste as barytes; but it is not poisonous; it is less soluble in water than barytes; it has the property of giving a red or purple colour to flame, for which purpose the nitrate or muriate may be dissolved in alcohol, or applied to the wick of a candle. The weight of the atom of strontites is deducible from the salts which it forms with the more common acids to be 46. Thus,

	Acid.		Base.		
Carbonate of strontites	29.2	+	70.8	per cent. as	19 : 46
Sulphate ———	42.5	+	57.5	———	34 : 46
Nitrate ———	45.2	+	54.8	———	38 : 46
Muriate ———	32.4	+	67.6	———	22 : 46

Dr. Hope, Pelletier, and Klaproth find 30 per cent. of acid in the carbonate. Klaproth, Clayfield, Henry, and Kirwan find 42 per cent. acid in the sulphate. Kirwan finds the crystallized nitrate to contain 31.07 acid, 36.21 base, and 32.72 water; which I presume denotes 1 atom of acid, 1 of base, and 5 of water; that is, 38 acid + 46 base + 40 water; this reduced, would give 30.6 acid, 37.1 base, and 32.3 water per cent. which very nearly agrees with his experience. Taking the dry salt, his results would give 46.2 acid, and 53.8 base. Vauquelin finds the nitrate to contain 48.4 acid, 47.6 base, and 4 water; but this constitution cannot be correct: Neither can Richter's analysis, which gives 51.4 acid and water, and 48.6 base.—Dry muriate of strontites, according to Kirwan, consists of 31 acid, and 69 base; but Vauquelin states 39 acid, and 61 base; the former, without doubt, is nearer the truth.

Hydrate of Strontites. When water is put to pure strontites, it becomes hot and swells, like lime and barytes, and falls into dry powder. This powder seems to be the first hydrate; whence, 46 parts of strontites will take 8 of water to form this combination; but if more water be added, the hydrate crystallizes. These crystals appear to be the 12th hydrate;

that is, they are constituted of 1 atom of strontites and 12 of water $= 46 + 96 = 142$, or 32 strontites $+$ 68 water per cent. agreeably to the experience of Dr. Hope. Water dissolves about $\frac{1}{160}$th of its weight of pure strontites in the temperature of 60°, or $\frac{1}{50}$th of its weight of the crystals; the specific gravity of the solution is nearly 1.008. But boiling water dissolves about half its weight of the crystals. Whence it appears that strontites is much less soluble than barytes, and much more soluble than lime. The specific gravity of the crystals of strontites is rightly determined by Hassenfratz to be nearly 1.46. Strontian water may be used for the same purposes as lime-water, or barytic water.

Strontites combines with most of the acids to form neutral salts. It also combines with sulphur and phosphorus.

5. *Alumine, or Argil.*

The earth denominated alumine, constitutes a great portion of common *clay*; but this last is a mixture of two or more earths with iron, &c., and therefore cannot be exhibited as pure alumine. The earth may be obtained pure from a common well known salt, called *alum*, which is constituted of sulphate of potash and sulphate of alumine combined together, with a portion of water. A quantity of alum is to be dissolved in 10 times its weight of water; to this a quantity of liquid ammonia is to be added; the sulphuric acid seizes the ammonia, and lets fall the alumine, which may be separated from the liquid by filtration; and then exposed to a red heat.

Alumine thus obtained is a fine white earth, spongy, and adhesive when moistened; it has neither taste nor smell; it is said to have the specific gravity, 2. When mixed with water, it forms a mass which is the basis of earthen ware, and capable of receiving any figure. In this case, by the application of great heat, it becomes excessively hard, and loses in part, or wholly, its adhesive quality. Pure alumine bears the highest heat of a furnace without undergoing any change.

Alumine does not form any known combination with oxygen, hydrogen, charcoal, sulphur, or phosphorus; but it combines with the alkalies, with most of the earths, and with several metallic oxides. It combines too with many of the acids, but forms in most cases uncrystallizable salts. It possesses a strong affinity for vegetable colouring matter, and hence its great importance in the arts of dyeing and printing, in which it is employed to fix the colour on the cloth.

The weight of an atom of alumine is not so easily determined as that of the preceding earths and alkalies; partly because the salts which it forms with the acids are not crystallizable, and partly because they have not had a proportionate share of attention paid to them. The only salt with alumine which has been carefully analyzed is the triple compound, or alum; an acquaintance with the constitution and properties of this salt is of great importance to its manufacturer, and to the various artists to whom it is of indispensible utility.

The experience of Chaptal, Vauquelin, and of Thenard and Roard (An. de Chimie, vol. 22, 50, and 59, or Nicholson's Journal, vol. 18) shews that the alum of all countries is very nearly the same in its constitution and qualities, that it contains 33 per cent. sulphuric acid, 11 or 12 alumine, 8 or 9 potash, and 47 water. All the authors I have mentioned do not agree, it is true, in these numbers; but the differences are more in appearance than reality. Vauquelin obtains 95 sulphate of barytes from 100 alum, but Thenard and Roard obtain 100. The last mentioned chemists adopt only 26 per cent. acid in sulphate of barytes; whereas it is now universally allowed there are about

33 per cent. acid in that salt. Mr. James Thomson, I am informed, finds nearly 100 per cent. sulphate of barytes. This result I adopt as the most correct, and it is also the most recent. Vauquelin finds $48\frac{1}{2}$ water in alum; this is more than is generally found, and accounts in some degree for his obtaining less sulphate of barytes. Chaptal finds 47 per cent. water in English alum, with which my experience accords. Vauquelin finds 10.5 alumine, Thenard and Roard, 12.5 per cent. Mr. Tennant of Glasgow, who favoured me with an analysis, finds 11.2 alumine in the alum manufactured there. This last chemist finds 15 per cent. sulphate of potash, which is the same as Thenard and Roard's nearly, 15.7. Now, as 34 acid + 42 potash, have been shewn to constitute 76 sulphate, 15 must contain 6.7 acid and 8.3 potash. Collecting these results then, it appears that alum may be said to consist of,

33 sulphuric acid.
11.7 alumine.
8.3 potash.
47 water.

100

Of the 33 sulphuric acid, it must be recollected that 6.7 parts belong to the potash; that is, $\frac{1}{5}$th of the whole; the remainder, or $\frac{4}{5}$ths, belong to the alumine. Hence, then, were there only 5 atoms of sulphuric acid in a molecule of alum, 1 atom would appertain to an atom of potash, and the other 4 atoms to as many of alumine, provided the acid and alumine unite one to one, which we are to presume till sufficient reason appear to the contrary. It should seem, then, that an atom of alum is constituted of one of sulphate of potash in the centre, and 4 atoms of sulphate of alumine around it, forming a square. But 33 — 6.7 = 26.3 acid to 11.7 alumine; and 26.3 : 11.7 : : 34 : 15, the weight of an atom of alumine. Dry alum must, therefore, be $5 \times 34 + 42 + 4 \times 15 = 272$; but as this is found combined with water in the state of common alum, it will be satisfactory to know how many atoms of water are attached to one atom of dry alum: for this purpose, we have 53 : 47 : : 272 : 241 = the weight of water; this, divided by 8 gives the number of atoms = 30. Hence, an atom of common alum consists of,

1 atom of sulphate of potash	= 76	= per cent.	15
4 atoms of sulphate of alumine	= 196	——	38
And 30 atoms of water.	= 240	——	47
	512		100

A saturated solution of alum in water, at the temperature 60°, is of the specific gravity 1.048, and is constituted of 1 atom of dry alum and 600 of water; or the alum has 20 times the quantity of water that the crystals contain. The specific gravity of alum itself is about 1.71; and by means of heat, solutions of it in water may be obtained of any inferior specific gravity; at least, I have had a solution, which, when hot, was 1.57.

Alumine does not combine with carbonic acid; but it combines with the nitric and muriatic acids; it would, therefore, be desirable that the weight of an atom of alumine should be investigated from these last combinations, as well as from the sulphate. No author that I know has given the proportion of elements in nitrate of alumine; and in muriate of alumine Bucholz determines equal parts of acid and base, and Wenzel 28 acid to 72 base; so that no confidence can be placed in them. I determined the proportions of these salts as follows: 100 grains of alum were dissolved in water; the alumine was precipitated by 156 measures, more or less, of test ammonia, (.97), care being taken that the aluminous solution was saturated with ammonia, and that none was superabundant; the liquid was then well agitated, and immediately divided into three equal portions. It was then found that each of these portions took 52 measures of the test acids; namely, the sulphuric, the nitric, and the muriatic respectively, to dissolve the floating alumine, and to clear the solutions which were afterwards found to be free from uncombined acids. Hence, the proportions of the salts are deduced as under:

	Acid.		Base.		
Sulphate of alumine	69.4	+	30.6	per cent. as	34 : 15
Nitrate —— ——	71.7	+	28.3	—— ——	38 : 15
Muriate —— ——	59.5	+	40.5	—— ——	22 : 15

It will be proper here to notice an opinion which Vauquelin supported in his essay in 1797, but which is not adverted to in his succeeding essay in 1804, nor in the one of Thenard and Roard in 1806: I mean the opinion that alum consists of the *supersulphate* of alumine and sulphate of potash. If this be true, then the atom of alumine must weigh 30, because 2 atoms of sulphuric acid unite to 1 of alumine. The opinion appears to me without support. When a solution of alum is put to the blue test, it changes it to red; but this is not a proof of excess of acid where the base of the salt has a strong affinity for colouring matter; there is probably a true decomposition of the salt, or perhaps the colouring matter forms a triple compound with the salt. That no uncombined acid accompanies alum is certain, because the least portion of alkali decomposes it. Besides, a red heat drives off half of the acid at least from supersalts; but alum bears a red heat without losing a sensible portion of acid. From the experiment related above, it appears that the sulphuric, the nitric, and the muriatic acid tests are of equal efficacy in saturating alumine. Are these all supersalts? If so, why does not half the acid in each case neutralize the earth, and form a simple salt?—But it is said if alumine be boiled in a solution of alum, the alumine combines with the alum, and falls down an insoluble, neutral salt. Vauquelin asserts he has made the experiment; but he mentions no proportions, nor does he point out the time requisite to produce the effect. With a view to this subject, I precipitated the alumine from a measure of saturated solution of alum at 60° (about 100 grains of alum) by the necessary quantity of ammonia; to this liquid, which was found neutral, still containing the alumine in suspension, I put another measure of the same solution of alum, and boiled the whole for 10 minutes in a glass vessel; it was then set aside to cool, and filtered; the liquid was not much diminished in specific gravity, and required nearly the same quantity of ammonia to saturate it, and afforded the same quantity of alumine as the first measure. Apprehending the sulphate of ammonia present might influence the result, I next put the dry pulverized alumine from 100 grains of alum into a solution of 100 grains of alum in water, and in another experiment the moist recently filtered alumine, and boiled the whole for 10 minutes; the water evaporated was restored, and the liquor filtered; it was of the same specific gravity as at first, tasted equally aluminous, and the precipitate collected and dried, weighed just the same as before. These facts lead me to doubt concerning the existence of this *alum saturated with its earth*, as the earlier chemists called it. But supposing the existence of a combination of sulphuric acid with twice the quantity of alumine, I know no reason why it should not be constituted of 1 atom of acid and 2 of alumine. Hence, I conclude the weight of an atom of alumine above stated is a fair deduction.

The French chemists seem to have proved that the presence of even a very small portion of sulphate of iron in alum is very injurious in some of its uses in dyeing, &c.

Hydrate of Alumine. Saussure, in the 52d vol. of the Journal de Physique, observes, that alumine is precipitated from its solution, in two very different states, according to circumstances; the one he calls *spongy*, and the other *gelatinous* alumine; they both retain 58 parts per cent. of water, when dried in common summer heat; the former parts with the whole of its water at a red heat; but the latter only loses 48 per cent. at the highest temperature. There may be some doubt as to the accuracy of these facts; but it would seem probable that alumine, at the ordinary temperature, retains 2 atoms of water, or 15 parts alumine hold 16 of water; this would allow 52 per cent. loss by a red heat. The subject deserves further attention.

6. *Silex.*

The earth denominated *silex*, is found abundantly in a great many stones; it is almost pure in *flint*, *rock crystal*, and others; but of stones in general it only constitutes a part, being found in combination with one or more of the other earths, or with metals, &c. It is also found in small particles in the form of white sand. The most distinguishing feature of this earth is its melting along with either of the fixed alkalies, and forming with them that beautiful and well known compound, glass. The specific gravity of flint and rock crystal is usually about 2.65. After being heated red hot for some time, flint may be pulverized in an iron mortar, and forms a white earth, which may be regarded as silex sufficiently pure for most purposes. It forms a harsh, gritty powder, which does not cohere nor form a paste with water like clay. It is insoluble in water in any sensible degree. It is infusible by heat, unless at an extremely high degree. To obtain silex in a pure state, a mixture of sulphuric acid and fluate of lime must be distilled in glass vessels, or along with pulverized flint, when superfluate of silex is produced in an elastic state; the gas may be received over water, on the surface of which a crust of fluate of silex is formed; this crust being removed by filtration or otherwise, the clear liquor is to be saturated with ammonia, when pure silex is thrown down. When dried in a red heat, it forms a fine white powder. The common mode prescribed to obtain pure silex gives pure glass, as will presently be explained. It is remarkable, that sulphuric acid, poured on fluate of silex, expels the fluoric acid in fumes, though it does not combine with the silex.

Silex combines with the two fixed alkalies, with most of the earths, and with metallic oxides; but with few of the acids immediately, except the fluoric; when joined to an alkali, it may be united to several of the acids, forming triple salts. It seems not to combine with oxygen, hydrogen, or the other combustibles, nor with ammonia.

The fixed alkalies may each be combined with silex in two proportions. In order to form glass, one part of silex and one of fine dry carbonate of soda may be mixed together; but if potash is used, then $1\frac{1}{2}$ parts will be required. If the other or soluble compound is wanted, then double the quantities of alkali must be used, or 2 parts of soda and 3 of potash. A strong red heat in each case is necessary to form a complete union of the principles; the fused mass gives out the carbonic acid of the alkalies, and when poured out immediately becomes glass; but when the double quantity of alkali is used, the glass is deliquescent, and may be completely dissolved in water. This last may be called *supersodiuretted* or *superpotasiuretted silex*, and the former *sodiuretted* or *potasiuretted silex*. When an acid is dropped into a solution of superpotasiuretted silex, a white precipitate is immediately formed, which is potasiuretted silex, or common glass, and not silex, as has hitherto been supposed. For, 1. The heated precipitate, I find, weighs about $\frac{2}{3}$ds of the red hot potasiuretted silex, whereas the silex is only about $\frac{1}{3}$d of the compound; 2. the acid requisite to throw down the precipitate, is only half of that which the alkali in the compound would require for its saturation; 3. the precipitate, dried in a moderate red heat, is fusible into glass by the blow-pipe; and, 4. as the acids do not take the alkali from glass, they ought not to take more alkali from superpotasiuretted silex than what would reduce it to common glass.

It is more difficult to find the weight of an atom of silex than that of any other of the previous earths, because it enters into combination with only one of the acids, and the proportions have not yet been ascertained. I have, however, succeeded pretty well by investigating its relations with potash, lime, and barytes. Having obtained a quantity of superpotasiuretted silex without any excess of alkali; that is, which afforded a precipitate with the least portion of acid (for if the alkali be in excess, acid may be added without any precipitation), I precipitated a given weight of the dried compound previously in water, by sulphuric acid in excess; the precipitate was heavy and bulky; after remaining on the filter for some time, it resembled a mass of over-boiled potatoe; the water being forced out by pressure, a white subtance remained, which easily left the filter, and when dried in a low red heat, left a harsh gritty powder, nearly $\frac{2}{3}$ds of the weight of the compound. Again, test sulphuric acid was slowly added to the solution, of a given weight of the dry compound in water; as soon as the mixture manifested acid to the test liquid, it was considered as saturated. The whole acid added was found to be sufficient to saturate a weight of pure alkali nearly equal to $\frac{1}{3}$d of that of the dry compound. These experiments rendered it obvious that only one half of the alkali was engaged by the acid, the other half remaining with the silex; and the conversion of the precipitate into glass by the blow-pipe confirmed the conclusion. It remained, then, to determine which of the two combinations of alkali and silex was the most simple. As a part of the alkali is easily drawn from one compound, and difficultly from the other, the former must be supposed two atoms of alkali to one of silex, and the latter one to one. From this it should seem, that the weight of an atom of silex is nearly the same as that of an atom of potash; and the near agreement of the specific gravities of these two bodies, is an argument in favour of the conclusion.

Superpotasiuretted silex exhibited remarkable results with lime and barytes. One hundred measures of the solution, containing 18 grains dry, were saturated with 5000 grains of lime water, containing 6 grains of lime; the precipitate, filtered and dried in a low red heat, was 19 grains. The residuary liquid required 27 grains of test muriatic acid to saturate it; whereas, the like quantity of lime water took 54 grains. Here, then, it appears that each atom of the superpotasiuretted silex must have been decomposed into one atom of potash, which remained in the liquid, and one atom of potasiuretted silex, which united to two atoms of lime, and the compound was precipitated. That the matter in the liquid was potash, and not lime, was proved by carbonic acid; and the test muriatic acid shewed that every atom of potash in the liquid took the place of two atoms of lime. The case was different with barytes. One hundred measures of the solution, containing 18 grains dry, were saturated with 850 measures of 1.0115 barytic water, containing 9 dry barytes. The residuary liquid took 28 test acid to saturate it, and the precipitate dried in a red heat was 20 grains. Here it is evident that one atom of barytes had detached one of potash from the compound, and taken its place; consequently, the residue of liquid required the same quantity of acid as the barytic water, and the precipitate was a triple compound of silex, potash, and barytes; one atom of each, consisting probably of 9 parts of barytes, $5\frac{1}{2}$ silex, and $5\frac{1}{2}$ potash.

Upon the whole, I am inclined to believe that one atom of silex weighs nearly 45 times that of hydrogen.

Silex combines with alumine by heat, and the compound forms hard infusible bodies, such as porcelain, earthen ware, bricks, &c.

7. *Yttria.*

This earth is found at Ytterby, in Sweden. It constitutes a portion of the mineral called *gadolinite*, first analyzed by Gadolin, and of that called *yttrotantalite*, both found in the same mine. The earth may be obtained by dissolving the pulverized mineral in a mixture of nitric and muriatic acids; the liquor poured off is then evaporated to dryness, the residuum dissolved in water. If ammonia be now added, the earth is precipitated. It is obtained in the form of a white powder, said to be of the specific gravity 4.84. It is infusible by heat, and insoluble in water: but it forms salts with several of the acids; and these salts have mostly a sweet taste, and are in some instances coloured. They resemble the metallic salts in many particulars. According to Klaproth, the hydrate of yttria, a dry powder, contains 31 per cent. water; this would imply that the atom of yttria weighs 18, 36, or 53, according as it is the first, second, or third hydrate; but he finds the carbonate of yttria to be 18 acid, 55 yttria, and 27 water: now, supposing the carbonate to be 1 atom of acid, 1 of earth, and 3 of water, and that the acid and water weigh 45, then the atom of earth is deduced to be 53; and this conclusion agrees with the preceding one, which supposes the hydrate to be the third. The great specific gravity of the earth countenances the notion of the atom being heavy; but we cannot rely upon the above determination till it is supported by more various experiments.

8. *Glucine.*

The earth called *glucine* (from the sweet-tasted salts which it forms with acids) is obtained chiefly from two minerals, the beryl and the emerald. These minerals are constituted of silex, alumine, and glucine; the two former being abstracted by the usual processes there remains the glucine, a soft white powder, adhering to the tongue, but without taste or smell, and infusible by heat. Its specific gravity is said to be 2.97. It is insoluble in water. This earth combines with the acids, with liquid fixed alkalies, and with liquid carbonate of ammonia. In the last case it resembles yttria, but is much more soluble than that earth in carbonate of ammonia Glucine has considerable resemblance in its properties both to alumine and yttria.

We have not data sufficient to find the weight of an atom of glucine; but from the experiments of Vauquelin on the carbonate of glucine (Annal. de Chimie, tom. 26, pages 160 and 172) it should seem to weigh nearly 30, or twice the weight of alumine. It is remarkable, too, that the analysis of the beryl, and of the emerald, give nearly the same quantity of alumine and glucine, which indicates that the weight of an atom of the latter is either equal to that of the former, or some multiple of it.

9. *Zircone.*

The *zircon* or *jargon*, and the *hyacinth*, are two precious stones found chiefly in Ceylon. These contain a peculiar earth which has received the name of zircone. It may be obtained thus: Let one part of zircon in powder, be fused with 6 parts of potash; then let the mass be diffused through a portion of water, which will dissolve the potash and its combinations, and leave a residuum. This residuum must be dissolved in muriatic acid, and potash must be added, which will precipitate the zircone. It is a fine white powder, insipid, and somewhat harsh to the feel. When violently heated, it is converted into a kind of porcelain, very hard, and of the specific gravity 4.35. Zircone is not soluble in water, but it retains $\frac{1}{3}$ or $\frac{1}{4}$ of its weight of water when dried in the air, and assumes the appearance of gum arabic. Zircone is not soluble in liquid alkalies, but it is in the alkaline carbonates; it adheres to several of the metallic oxides. Zircone unites with acids, and forms with them salts, many of which are insoluble in water, but others are very soluble. They have an astringent taste, resembling some of the metallic salts.

As the salts of zircone have not yet been formed with sufficient care to ascertain the ratio of their constituent principles, we can not exactly determine the weight of an atom of this earth. Vauquelin finds 44 carbonic acid and water and 56 zircone in carbonate of zircone; but, unfortunately, he has not given the acid separately from the water. Allowing the accuracy of the above, and supposing the carbonate to contain 1 atom of water, the weight of an atom of zircone will be 34; but if we suppose 2 atoms of water, then the atom of earth comes out 45. This last I judge to be nearest the truth. It is remarkable, that the hyacinth contains 32 parts of silex and 64 of zircone, which, according to the above conclusion, corresponds to 1 atom of silex and 2 of zircone, a constitution by no means improbable. Upon this principle, the gummy hydrate above mentioned, may be 2 atoms of water and 1 of zircone, or 16 water + 45 zircone.

END OF PART SECOND.

EXPLANATION OF PLATES.

PLATE 5. Exhibits the various symbols devised to represent the simple and compound elements; they are nearly the same as in plate 4, only extended and corrected: they will be found to agree with the results obtained in the preceding pages.

Fig.	Simple.	Wt.	Fig.		Wt.
1.	Oxygen	7	12.	Iron	50
2	Hydrogen	1	13.	Nickel	25 ? 50 ?
3.	Azote	5	14.	Tin	50
4.	Carbone	5.4	15.	Lead	95
5.	Sulphur	13	16.	Zinc	56
6.	Phosphorus	9	17.	Bismuth	68 ?
7	Gold	140 ?	18.	Antimony	40
8.	Platina	100 ?	19.	Arsenic	42 ?
9.	Silver	100	20.	Cobalt	55 ?
10.	Mercury	167	21.	Manganese	40 ?
11.	Copper	56	22.	Uranium	60 ?

Fig.		Wt.	Fig.		Wt.
23.	Tungsten	56 ?	41.	Nitrous gas	12
24.	Titanium	40 ?	42.	Nitrous oxide	17
25.	Cerium	45 ?	43.	Nitric acid	19
26.	Potash	42	44.	Oxynitric acid	26
27.	Soda	28	45.	Nitrous acid	31
28.	Lime	24	46.	Carbonic oxide	12.4
29.	Magnesia	17	47.	Carbonic acid	19.4
30.	Barytes	68	48.	Sulphurous oxide	20
31.	Strontites	46	49.	Sulphurous acid	27
32.	Alumine	15	50.	Sulphuric acid	34
33.	Silex	45	51.	Phosphorous acid	32
34.	Yttria	53	52.	Phosphoric acid	23
35.	Glucine	30	53.	Ammonia	6
36.	Zircone	45	54.	Olefiant gas	6.4
			55.	Carburetted hyd.	7.4
	Compound:		56.	Sulphuret. hydr.	14
37.	Water	8	57.	Supersulph. hydr.	27
38.	Fluoric acid	15	58.	Phosphuret. hydr.	10
39.	Muriatic acid	22	59.	Phosphur. sulph.	22
40.	Oxymuriatic acid	29	60.	Superphos. sulph.	31

PLATE 6. Symbols of compound elements (continued from Plate 5.)

Fig.		Wt.	Fig.		Wt.
1.	Hydrate of potash	50	16.	Muriate of barytes	90
2.	Potasium, or hydruret of potash	43	17.	Sulphate of alumine	49
3.	Carbonate of potash	61	18.	Nitrate of alumine	53
4.	Hydrate of soda	36	19.	Muriate of alumine	37
5.	Sodium, or hydruret of soda	29	20.	Alum	272
6.	Carbonate of soda	47	21.	Potasiuretted silex, or glass	87
7.	Hydrate of lime	32	22.	Superpotasiuretted silex	129
8.	Carbonate of lime	43	23.	Potash, silex, & lime	135
9.	Sulphate of lime	58	24.	Potash, silex, & barytes	155
10.	Nitrate of lime	62	25.	Fluate of silex	60
11.	Muriate of lime	46	26.	Subpotasiuretted * ammonia	54
12.	Hydrate of barytes	76	27.	Oxymuriate of olefiant gas	41
13.	Carbonate of barytes	87			
14.	Sulphate of barytes	102			
15.	Nitrate of barytes	106			

* The olive coloured substance obtained by heating potasium in ammoniacal gas, by Gay Lussac and Thenard, Davy, &c.

PLATE 7. Fig. 1, 2, and 3. represent profile views of the disposition and arrangement of particles constituting elastic fluids, both simple and compound, but not mixed; it would be difficult to convey an adequate idea of the last case, agreeably to the principles maintained, page 190.—The principle may, however, be elucidated by the succeeding figures.

Fig. 4. is the representation of 4 particles of azote with their elastic atmospheres, marked by rays emanating from the solid central atom; these rays being exactly alike in all the 4 particles, can meet each other, and maintain an equilibrium.

Fig, 5. represents 2 atoms of hydrogen drawn in due proportion to those of azote, and coming in contact with them; it is obvious that the atoms of hydrogen can apply one to the other with facility, but can not apply to those of azote, by reason of the rays not meeting each other in like circumstances; hence, the cause of the intestine motion which takes place on the mixture of elastic fluids, till the exterior particles come to press on something solid.

PLATE 8. The first 16 figures represent the atoms of different elastic fluids, drawn in the centres of squares of different magnitude, so as to be proportionate to the diameters of the atoms as they have been herein determined. Fig. 1. is the largest; and they gradually decrease to fig. 16, which is the smallest; namely, as under,

Fig.		Fig.	
1.	Superfluate of silex	9.	Oxymuriatic acid
2.	Muriatic acid	10.	Nitrous gas
3.	Carbonic oxide	11.	Sulphurous acid
4.	Carbonic acid	12.	Nitrous oxide
5.	Sulphuretted hydrogen	13.	Ammonia
6.	Phosphuretted hydrogen	14	Olefiant gas
7.	Hydrogen	15.	Oxygen
8.	Carburetted hydrogen	16.	Azote.

Fig. 17. exhibits curve lines, by which the boiling point of liquid solutions of nitric and muriatic acid, and of ammonia, of any strength, may be determined. They are representations of the results contained in the preceding tables relative to these articles. If any point be taken in one of the curves, and a horizontal line be traced to the margin, the strength per cent. by weight of the liquid will be shewn; and if a perpendicular line be traced to the top, the temperature at which the liquid of that strength boils in the open air will be found.

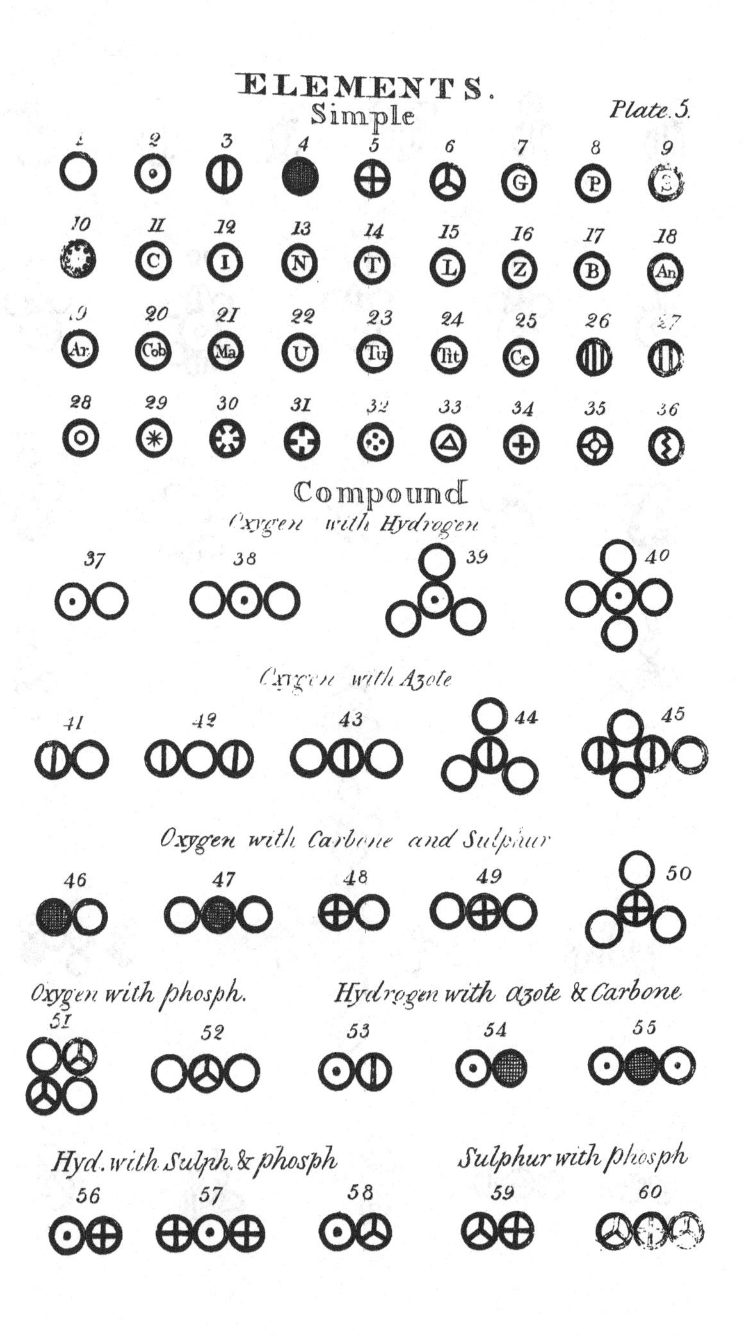
ELEMENTS.
Simple
Plate. 5.
1 2 3 4 5 6 7 8 9
G P
10 11 12 13 14 15 16 17 18
C I N T L Z B An
19 20 21 22 23 24 25 26 27
Ar Cob Ma U Tu Tit Ce
28 29 30 31 32 33 34 35 36
Compound
Oxygen with Hydrogen
37 38 39 40
Oxygen with Azote
41 42 43 44 45
Oxygen with Carbone and Sulphur
46 47 48 49 50
Oxygen with phosph.
51 52
Hydrogen with azote & Carbone
53 54 55
Hyd. with Sulph. & phosph
56 57 58
Sulphur with phosph
59 60

Plate 6.

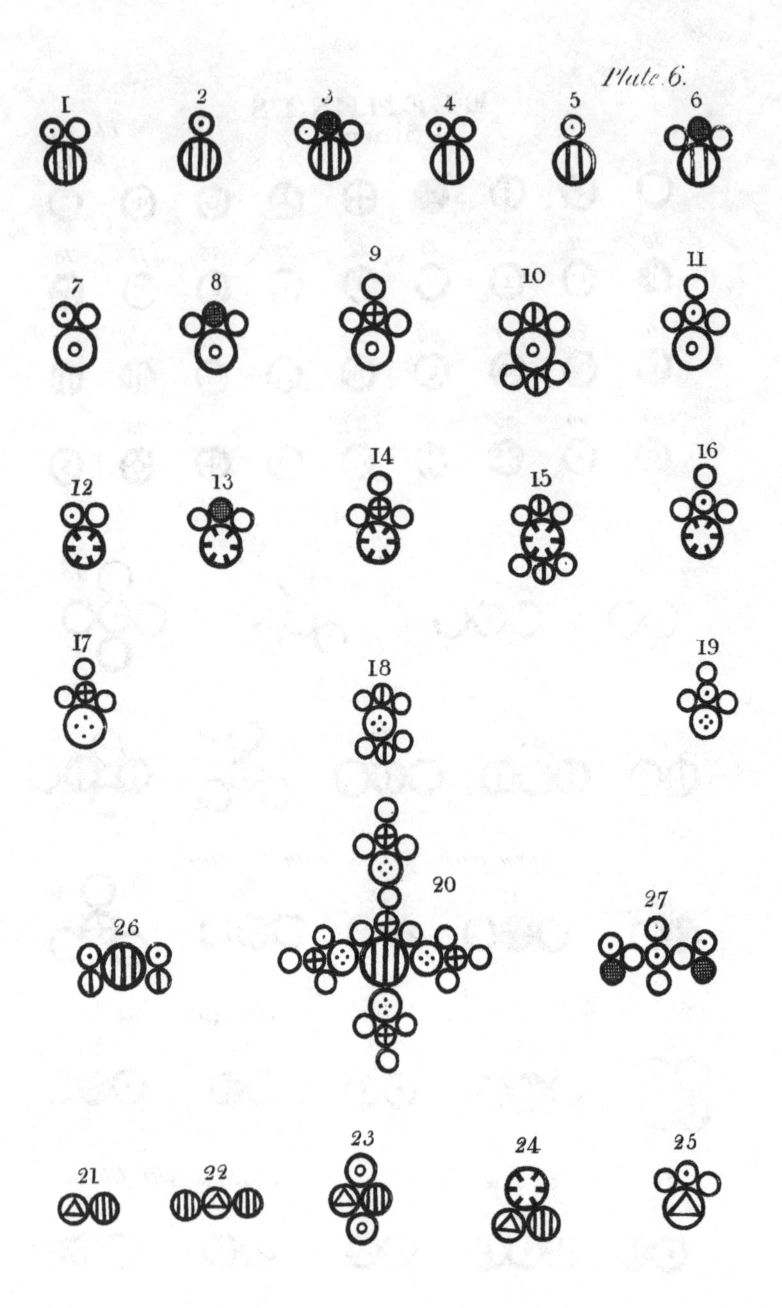

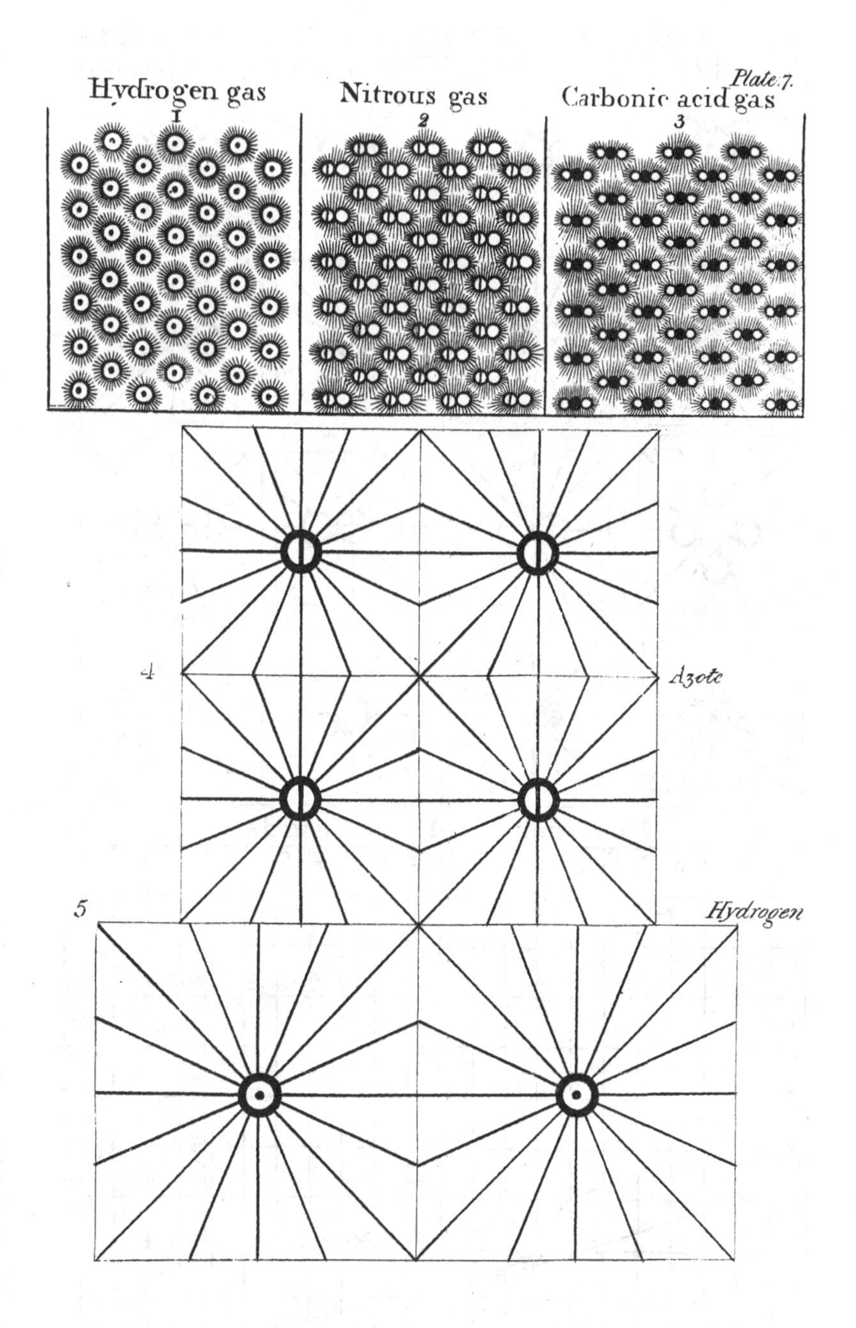
Plate 7.
Hydrogen gas
Nitrous gas
Carbonic acid gas
1
2
3
4
Azote
5
Hydrogen

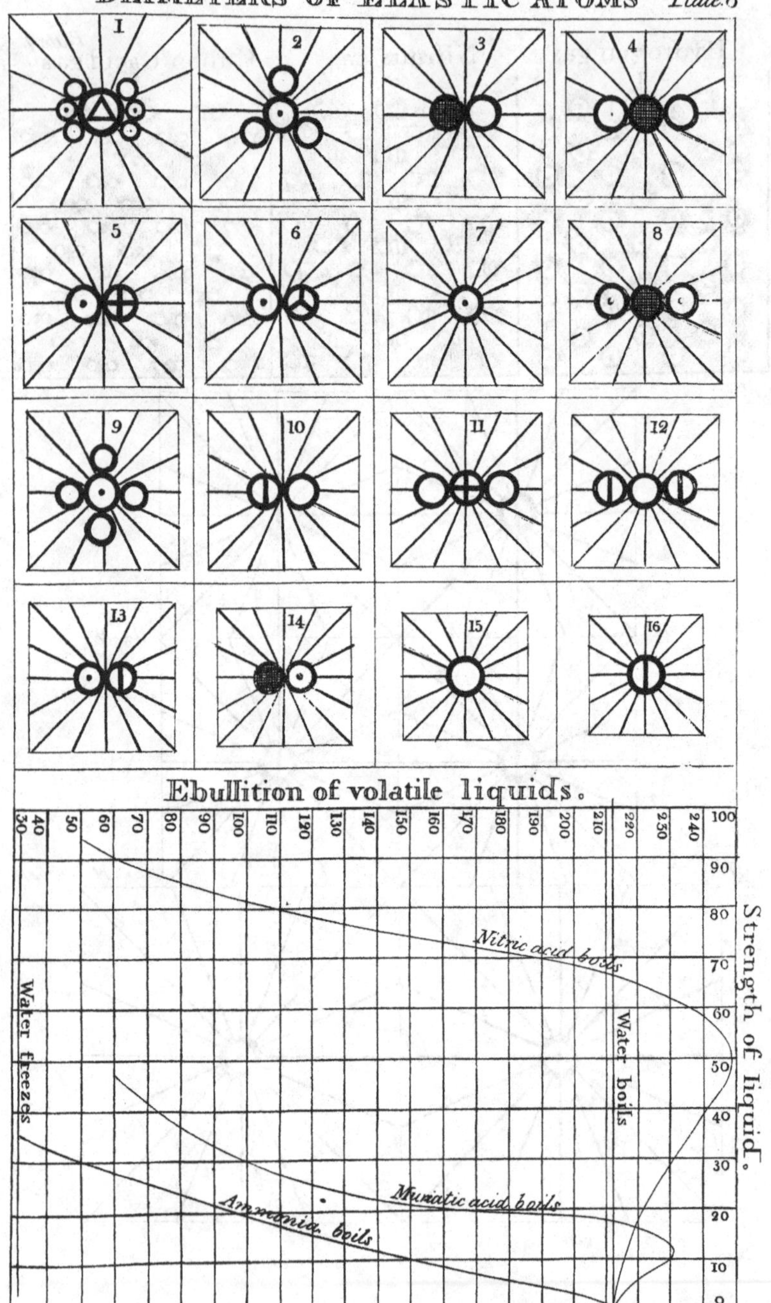
DIAMETERS OF ELASTIC ATOMS
Plate. 8
1
2
3
4
5
6
7
8
9
10
11
12
13
14
15
16
Ebullition of volatile liquids.
30
40
50
60
70
80
90
100
110
120
130
140
150
160
170
180
190
200
210
220
230
240
100
90
80
70
60
50
40
30
20
10
0
Nitric acid boils
Water boils
Water freezes
Muriatic acid boils
Ammonia boils
Strength of liquid.

APPENDIX.

As it is nearly two years since the printing of this second part commenced, it may be expected that in the rapid progress of chemical investigation, some addition has, in the interval, been made to the stock of facts and observations relating to the more early subjects herein discussed. The ground upon which I determine the weights of the ultimate particles of the metals, has not yet been entered upon. This will occupy a leading place in a second volume, when the metallic oxides and sulphurets come to be considered. It will be observed, that I have seen reason to change some of the metallic weights which were given in the first part; and it is probable, that in our future investigations these may be again changed; this will depend upon the precision with which the proportions of the elements of the metallic oxides, sulphurets and salts, shall be obtained. The identity of tantalium and columbium seems to have been ascertained by

Dr. Wollaston. Mr. Davy, and the French chemists Gay Lussac and Thenard, have furnished a number of facts and observations on various subjects, resulting from their application of the new metals, potasium and sodium, and Voltaic electricity, to chemical investigations. When the mind is ardently engaged in prosecuting experimental enquiries, of a new and extraordinary kind, it is not to be expected that new theoretic views can be examined in all their relations, and formed so as to be consistent with all the well known and established facts of chemistry; nor that the facts themselves can be ascertained with that precision which long experience, an acquaintance with the instruments, and the defects to which they are liable, and a comparison of like observations made by different persons, are calculated to produce. This may appear to be a sufficient apology for the differences observed in the results of the above celebrated chemists, and for the opposition, and sometimes extravagance, of their views.

All the phenomena of combustion are exhibited by heating potasium in fluoric acid gas (superfluate of silex); though this would seem to intimate that the gas contains oxygen, yet, as Mr. Davy properly observes, heat and light are merely the results of the intense agency of combination. It is remarkable that hydrogen is given out, yet not so much as would be given by the action of potasium on water; it is variable, and amounts generally to less than $\frac{1}{4}$th of that quantity. Mr. Davy and the French chemists agree in the belief of a decomposition of the acid; but it is doubtful whether the hydrogen is from the potasium or the acid. The fact, I have observed, page 285, of the diminution of a mixture of hydrogen and fluoric acid gas by electricity, is one of the strongest in favour of the notion that the acid gas contains oxygen.

Muriatic acid has been a great object of investigation. Mr. Davy's ideas on this subject, in his Electrochemical Researches, 1808, were, that the acid gas contains water in a combined state; or, to use my own phraseology, that an atom of real muriatic acid combined with one of water, formed one of the acid gas; hence, in burning potasium in the gas, the potasium decomposed the water, the hydrogen was liberated, and the oxygen joined to the potasium to form potash, with which the real or dry acid immediately united. This conclusion was plausible; but it was truly astonishing to see the French chemists draw the same conclusion from their views of the subject. They should have viewed muriatic acid gas as the pure acid, which combined with the potash of the potasium, and liberated its hydrogen. Mr. Davy has recently written an essay on the oxymuriatic and muriatic acids, with a copy of which he has just favoured me ; in this, he discards his former opinion of the gaseous combination of acid and water, and adopts another, that muriatic acid gas is a pure elastic fluid, resulting from the union of hydrogen with oxymuriatic acid, which last he conceives to be a simple substance. This notion agrees so far with mine, as to make hydrogen the base of muriatic acid ; but I cannot adopt his constitution of the acid. Mr. Davy now considers the hydrogen liberated, by the combustion of potasium in muriatic acid gas, as proceeding from the decomposed acid, and the new compound an *oxymuriate* of *potasium*. The explanation I prefer is, that the hydrogen proceeds from the potasium, and the undecomposed acid gas unites to the potash.

As to oxymuriatic acid, Gay Lussac and Thenard have reported some very striking and unexpected properties of it which they have discovered. They assert, that dry oxymuriatic acid gas was not decomposed by sulphurous acid gas, nitrous oxide, carbonic oxide, nor even nitrous gas, when these were dry; but that it was immediately decomposed by them if water was present. These *may* appear to them to be facts; but certainly they are too important, and some of them too difficultly ascertained, to be believed merely upon the assertion of any one. By what means were they found? What was the structure of the apparatus, the quantity of gases operated upon, the time they were allowed to be in contact, the means employed to investigate the results, &c. &c.? To answer all these enquiries satisfactorily, would require a volume in detail; yet, Gay Lussac and Thenard have not said one word. Now, we know that the facts respecting the mixtures of these gases over water, are *not* as above stated. Mr. Davy observes, (Researches, page 250) that "oxygenated muriatic acid and nitrous oxide "were mingled in a water apparatus; there "was a slight appearance of condensation; "but this was most probably owing to absorp-"tion by the water; on agitation, the oxy-"genated muriatic acid was absorbed, and the "greater part of the nitrous oxide remained un-"altered." I have repeatedly mixed carbonic oxide and nitrous gas with oxymuriatic acid in a water apparatus; the former mixture exhibits no signs of chemical union for several seconds; afterwards, if the sun shine upon it, chemical action commences, and continues somewhat slower than that of oxygen and nitrous gas; but if the mixture be put in the dark, it will remain for days, I believe, without any change. The latter mixture, or nitrous gas and oxymuriatic acid, in equal measures, over water, produces an instantaneous union, much more rapid than that of oxygen and nitrous gas, and which to all appearance seems independent upon the water. Now, if these simple experiments give such different results in different hands, what may we expect of the complex experiments, where the gases are previously dried, and then mixed in vessels quite free from mercury and water, and lastly examined after such mixture has taken place, regard being still had to the effects which mercury and water have, or are supposed to have, upon such mixtures?

Mr. Davy has given several experiments to shew that oxymuriatic acid combines with hydrogen to produce muriatic acid; but none of them appears to me decisive. When equal measures of hydrogen and oxymuriatic acid were introduced into an exhausted vessel, and fired by an electric spark, the result was a slight vapour, and a condensation of $\frac{1}{10}$ to $\frac{1}{20}$ of the volume, the gas remaining being muriatic acid. This fact, if it can be relied upon, is favourable to the notion it is to support; I should have expected a condensation of $\frac{1}{3}$ or $\frac{1}{4}$ of the total volume on the common hypothesis; if the author had described the apparatus and quantity of gases submitted to the experiment, with the mode of determining the quantity and quality of the residual gas, it would have assisted in any future enquiry on the subject; it is certainly an important experiment. Mr. Davy allows the hyperoxymuriate of potash to abound with oxygen. He supposes the oxygen to be attracted by the potasium, or the potash, rather than by the oxymuriatic acid. The facts appear to me to draw the other way much more powerfully. We find oxymuriatie acid in conjunction with much oxygen, in several other salts, but potash no where, except when joined to this acid.

Some observations on nitric acid, and the other compounds of azote and oxygen, have been made by Gay Lussac, in the 2d vol. of the Memoires d'Arcueil. He contends that one *measure* of oxygenous gas unites to two *measures* of nitrous gas to form nitric acid, and to three measures to form nitrous acid. Now

I have shewn, page 328, that 1 measure of oxygen may be combined with 1.3 of nitrous gas, or with 3.5, or with any intermediate quantity whatever, according to circumstances, which he seems to allow; what, then, is the nature of the combinations below 2, and above 3, of nitrous gas? No answer is given to this; but the opinion is founded upon an hypothesis that all elastic fluids combine in equal measures, or in measures that have some simple relation one to another, as 1 to 2, 1 to 3, 2 to 3, &c. In fact, his notion of measures is analogous to mine of atoms; and if it could be proved that all elastic fluids have the same number of atoms in the same volume, or numbers that are as 1, 2, 3, &c. the two hypotheses would be the same, except that mine is universal, and his applies only to elastic fluids. Gay Lussac could not but see (page 188, Part 1. of this work) that a similar hypothesis had been entertained by me, and abandoned as untenable; however, as he has revived the notion, I shall make a few observations upon it, though I do not doubt but he will soon see its inadequacy.

Nitrous gas is, according to Gay Lussac, constituted of equal measures of azote and oxygen, which, when combined, occupy the same volume as when free. He quotes Davy, who found 44.05 azote, and 55.95 oxygen by weight, in nitrous gas. He converts these into volumes, and finds them after the rate of 100 azote to 108.9 oxygen. There is, however, a mistake in this; if properly reduced, it gives 100 azote to 112 oxygen, taking the specific gravities according to Biot and Arago. But that Davy has overrated the oxygen 12 per cent. he shews by burning potasium in nitrous gas, when 100 measures afforded just 50 of azote. The degree of purity of the nitrous gas, and the particulars of the experiment, are not mentioned. This one result is to stand against the mean of three experiments of Davy, (see page 318) and may or may not be more correct, as hereafter shall appear. Dr. Henry's analysis of ammonia embraces that of nitrous gas also; he finds 100 measures of ammonia require 120 of nitrous gas for their saturation. Now this will apply to Gay Lussac's theory in a very direct manner; for, according to him, ammonia is formed of 1 measure of azote and 3 of hydrogen, condensed into a volume of 2; it follows, then, that 100 ammonia require 75 oxygen to saturate the hydrogen; hence, 120 nitrous gas should contain 75 oxygen, or 100 should contain 62.5, instead of 50. Here either the theory of Gay Lussac, or the experience of Dr. Henry, must give results wide of the truth. In regard to ammonia too, it may farther be added, that neither is the rate of azote to hydrogen 1 to 3, nor is the volume of ammonia doubled by decomposition, according to the experiments of Berthollet, Davy, and Henry, made with the most scrupulous attention to accuracy, to which may be added my own.—There is another point of view in which this theory of Gay Lussac is unfortunate, in regard to ammonia and nitrous gas; 1 measure of azote with 3 of hydrogen, forms 2 of ammonia; and 1 measure of azote with 1 of oxygen, forms 2 of nitrous gas: now, according to a well established principle in chemistry, 1 measure of oxygen ought to combine with 3 of hydrogen, or with one half as much, or twice as much; but no one of these combinations takes place. If Gay Lussac adopt my conclusions, namely, that 100 measures of azote require about 250 hydrogen to form ammonia (page 433), and that 100 azote require about 120 oxygen to form nitrous gas (page 331), he will perceive that the hydrogen of the former would unite to the oxygen of the latter, and form water, leaving no excess of either, further than the unavoidable errors of experiments might produce; and thus the great chemical law would be preserved. The truth is, I believe, that gases do not unite in equal or exact measures in any one instance; when they appear to do so, it is owing to the inaccuracy of our experiments. In no case, perhaps, is there a nearer approach to mathematical exactness, than in that of 1 measure of oxygen to 2 of hydrogen; but here, the most exact experiments I have ever made, gave 1.97 hydrogen to 1 oxygen.

I shall close this subject, by presenting two tables of the elements of elastic fluids; they are collected principally from the results already given in detail, with a few small alterations or corrections; the utility of them to practical chemistry will be readily recognised.

Tables of the elements of elastic fluids; at a mean temperature and pressure.

(TABLE 1.)

Names of the gases.	Wt. of an atom.	Wt. of 100 cubic inch. grs.	Specific gravity.	Diameter of an atom	No. of atoms in a given volume.
Atmospheric air	——	31	1.00	——	——
Hydrogen	1	2.5	.08	1.000	1000
Oxygen	7	34	1.10	.794	2000
Azote	5	30.2	.97	.747	2400
Muriatic acid	22	39.5	1.24	1.12	700
Ammonia	6	18.6	.60	.909	1330
Oxymur. acid	29	76	2.46	.981	1060
Nitrous gas	12	32.2	1.04	.980	1060
Nitrous oxide	17	50	1.60	.947	1180
Carbonic oxide	12.4	29	.94	1.020	940
Carbonic acid	19.4	47	1.52	1.00	1000
Sulphurous acid	27	71	2.30	.95	1170
Olefiant gas	6.4	29.5	.95	.81	1890
Carburetted hyd.	7.4	18.6	.60	1.00	1000
Sulphureted hyd.	14	36	1.16	1.00	1000
Phosphur. hyd.	10	26	.84	1.00	1000
Superflu. of silex	75	130	4.20	1.15	658

(TABLE 2.)

Proportions of the constituent principles of compound gases.

Names of the compound gases.	Constituent principles of 100 measures of the compound gases. Measures.	Measures.	Constituent principles of 100 weight of the compound gases.	
Ammon. gas	52 azote	+ 133 hyd.	83 azote	+ 17 hyd.
Water	100 oxyg.	+ 200 hyd.*	87 oxy.	+ 12.5 hyd.
Nitrous gas	46 azote	+ 55 oxyg.	42 azote	+ 58 oxygen
Nitr. oxide	99 azote	+ 58 oxyg.	59 azote	+ 41 oxygen
Nitric acid	180 nit. gas	+ 100 oxy.	27 azote	+ 73 oxy.
Nitrous acid	360 nit. gas	+ 100 oxy.	33 azote	+ 67 oxy.
Oxym. acid	150 mur. acid	+ 50 oxy.	76 mur. acid	+ 24 oxy.
Sulphs. acid	100 oxygen	+ sulphur	52 oxy.	+ 48 sulphur
Sulphc. acid	100 sulphs. acid	+ 50 oxy.	79½ sul. acid	+ 20½ oxy.
Carb. oxide	47 oxy.	+ charcoal	55 oxy.	+ 45 charc.
Carb. acid	100 oxy.	+ charcoal	72 oxy.	+ 28 charc.
Carbur. hyd.	200 hydr.	+ 1 part char.	27 hyd.	+ 73 charc.
Olefiant gas	200 hydr.	+ 2 parts ch.	15 hyd.	+ 85 charc.
Sulph. hyd.	100 hydr.	+ sulphur	7 hyd.	+ 93 sulph.
Mur. of am.	100 mur. acid	+ 100 am. g.	65 mur. acid	+ 35 am. gas
Carb. of am.	100 carb. acid	+ 80 am. g.	76 carb. acid	+ 24 am. gas
Subc. of am.	100 carb. acid	+ 160 am. g.	61 carb. acid	+ 39 am. gas

* I believe 197 is nearer the truth.

Printed in the United States
By Bookmasters